QUANTUM SCALING IN MANY-BODY SYSTEMS

An Approach to Quantum Phase Transitions

Quantum phase transitions are strongly relevant in a number of fields, ranging from condensed matter to cold atom physics and quantum field theory. This book, now in its second edition, approaches the problem of quantum phase transitions from a new and unifying perspective. Topics addressed include the concepts of scale and time invariance and their significance for quantum criticality, as well as brand new chapters on superfluid and superconductor quantum critical points, and quantum first-order transitions. The renormalisation group in real and momentum space is also established as the proper language to describe the behaviour of systems close to a quantum phase transition. These phenomena introduce a number of theoretical challenges which are of major importance for driving new experiments. Being strongly motivated and oriented towards understanding experimental results, this is an excellent text for graduates as well as theorists, experimentalists and those with an interest in quantum criticality.

MUCIO CONTINENTINO is a Professor at Centro Brasileiro de Pesquisas Físicas, Rio de Janeiro, and is a member of the Brazilian Academy of Sciences and Fellow of Conselho Nacional de Desenvolvimento Cientifico e Tecnológico-CNPq. His research interests are in condensed matter physics, with emphasis on the areas of strongly correlated electronic systems and quantum criticality.

QUANTUM SCALING IN MANY-BODY SYSTEMS

An Approach to Quantum Phase Transitions

MUCIO CONTINENTINO

Centro Brasileiro de Pesquisas Físicas, Brazil

CAMBRIDGE
UNIVERSITY PRESS

CAMBRIDGE
UNIVERSITY PRESS

University Printing House, Cambridge CB2 8BS, United Kingdom

One Liberty Plaza, 20th Floor, New York, NY 10006, USA

477 Williamstown Road, Port Melbourne, VIC 3207, Australia

4843/24, 2nd Floor, Ansari Road, Daryaganj, Delhi – 110002, India

79 Anson Road, #06–04/06, Singapore 079906

Cambridge University Press is part of the University of Cambridge.

It furthers the University's mission by disseminating knowledge in the pursuit of education, learning, and research at the highest international levels of excellence.

www.cambridge.org
Information on this title: www.cambridge.org/9781107150256
DOI: 10.1017/9781316576854

© Mucio Continentino 2017

First published 2017

Printed in the United States of America by Sheridan Books, Inc.

A catalogue record for this publication is available from the British Library.

Library of Congress Cataloging-in-Publication Data
Names: Continentino, Mucio A. (Mucio Amado)
Title: Quantum scaling in many-body systems / Mucio Continentino, Centro Brasileiro de Pesquisas Físicas, Brazil.
Description: Cambridge : Cambridge University Press, [2017] | Includes bibliographical references and index.
Identifiers: LCCN 2016053992 | ISBN 9781107150256
Subjects: LCSH: Phase transformations (Statistical physics) | Quantum theory.
Classification: LCC QC175.16.P5 C68 2017 | DDC 530.4/74–dc23
LC record available at https://lccn.loc.gov/2016053992

ISBN 978-1-107-15025-6 Hardback

to Sonia, Ilana and Marcelo

Contents

Preface

Quantum phase transitions remain an exciting and vigorous scientific subject. This field has experienced continuous theoretical progress. Equally important is the intense experimental work the field has engendered, including the discovery of new materials exhibiting different forms of quantum criticality that in turn further advance the theory. Thus, the new, enlarged edition of this book is more than timely.

The layout maintains the general structure of the first edition. New material has been added to the original chapters, and some of them have been modified, even substantially – such as the one on first-order quantum phase transitions. Moreover, I have written entirely new chapters on the study of topological quantum phase transitions and superconducting quantum critical points.

The book emphasises a general scaling approach that considers the change of parameters of a system under length and timescales close to quantum criticality. This provides a unifying view to such diverse phenomena as zero temperature magnetic transitions, metal-insulator, topological, superconductor and first-order quantum phase transitions. As in the previous edition, the renormalisation group with its concepts such as fixed points associated with scale invariance, crossover, and relevant and irrelevant interactions, provides the mathematical framework for the scaling ideas. It is the basic tool used to study, understand and describe quantum critical phenomena, even when the usual Landau paradigm fails, as for topological quantum phase transitions.

I would like to thank the many colleagues with whom I have been discussing and learning the subjects of this book. Among them, G. Aeppli, Elisa Baggio-Saitovitch, Sergey Bud'ko, Premala Chandra, Piers Coleman, Mauricio D. Coutinho-Filho, Alvaro Ferraz, Zach Fisk, Jacques Flouquet, Magda Fontes, Alex Lacerda, Mireille Lavagna, Hilbert v. Löhneysen, Andy J. Millis, Luiz Nunes de Oliveira, Pascoal Pagliuso, Silke Paschen, Aline Ramires, Daniel Reyes-Lopez, Peter Riseborough, Subir Sachdev, Andre Schuwartz, Qimiao Si, Frank Steglich, Manuel Nunez-Regueiro, Jean-Louis Tholence, Joe Thompson, Amos Troper and the lates Bruno Elschner and Bernard Coqblin. My sincere thanks to Alfredo Ozorio de Almeida, Claudine Lacroix, Eduardo Marino, Eduardo Miranda, Enzo Granato, Heron Caldas, Jose A. Helayël-Neto, Jose Abel Hoyos-Neto, Pedro Sacramento and Tharnier Puel de Oliveira for enlightening discussions and several suggestions to improve the new manuscript. The author, however, is solely responsible for any mistakes or omissions in the book.

I would like also to express my recognition of the encouragement of my wife Sonia, my daughter Ilana and my son Marcelo.

My sincere thanks to my editor, Simon Capelin, for the encouragement and support from the beginning to the completion of this book. Last but not least, I would like to thank the Brazilian agencies, CNPq, CAPES and FAPERJ for partial financial support.

1

Scaling Theory of Quantum Critical Phenomena

1.1 Quantum Phase Transitions

Quantum phase transitions, in contrast to temperature-driven critical phenomena, occur due to a competition between different parameters describing the basic interactions of the system. Their specific feature is the quantum character of the critical fluctuations. This implies, through the uncertainty principle, that energy fluctuations and time are coupled. Then at zero temperature time plays a crucial and fundamental role, the static properties being entangled with the dynamics (Continentino, 1994a; Sondhi *et al.*, 1997; Sachdev, 1999). In this book we are mainly interested in quantum phase transitions that occur in electronic many-body systems and how scaling concepts can be useful to understand their properties close to these transitions (Continentino, 1994a). Even though a similar approach has been used in the case of interacting bosons (Fisher *et al.*, 1989) the fermionic problem has its own idiosyncrasies and difficulties. For example, there is no natural order parameter associated with the localisation transition in the electronic case, while bosons at zero temperature are either localised or superfluid so that the superfluid order parameter can be used to distinguish between both phases.

If the results of the study of quantum phase transitions were restricted to zero temperature, this would be an interesting but purely academic area of research. What is really exciting about this subject is the effect of quantum critical points (QCP) in the finite temperature phase diagram of actual physical systems, even far away from the QCP (Freitas, 2015). As we will show, there is a special line in this phase diagram, the *quantum critical trajectory*, where the temperature dependence of the thermodynamic and transport properties is governed by the quantum critical exponents, i.e. those associated with the QCP. This and the observed crossover effects induced by temperature in the non-critical side of the phase diagram of a material with a QCP are sufficient to make the study of quantum phase transitions an inevitable subject.

1

We start this chapter by introducing the scaling theory of quantum critical phenomena, since scaling concepts are used throughout this book. We define the critical exponents and emphasise the special role played by the dynamic exponent in quantum phase transitions. For the purpose of making the scaling approach more concrete, we consider the simplest model exhibiting a quantum phase transition, namely the Ising model in a transverse field (Pfeuty, 1970). In the following chapters the basic tools to investigate critical phenomena are introduced, namely Landau theory and the renormalisation group in its real and momentum space versions. Next, we discuss the heavy fermion problem as a non-conventional but rich example of the application of the ideas of quantum criticality to a strongly correlated fermionic material. Heavy fermions are electronic systems with unstable f-shell elements that are close to a magnetic instability and have a huge thermal effective mass. The proximity of these systems to a magnetic quantum critical point, together with the ease of varying the distance to this QCP with external pressure, makes them ideal case studies for using the approach developed in this book. A scaling theory bearing in the competition between magnetism and Kondo effect is developed and applied to well known heavy fermion materials allowing the appreciation of its usefulness and limitations. These systems played an important role in the genesis of the new paradigm of quantum criticality as applied to strongly correlated materials (Continentino *et al.*, 1989).

In the following chapters the scaling approach is used to study the problem of metal-insulator transitions. These can be classified in two main categories: transitions where electronic correlations are not important and the correlation-driven Mott transition (Mott, 1974). Although the mechanisms that drive these transitions are distinct, they have in common the absence both of an order parameter and its conjugate field to distinguish the different phases. The concepts of scaling and critical exponents associated with quantum criticality are still useful to approach these transitions, even if all critical exponents cannot be properly defined.

Although recent theoretical developments, such as the dynamical mean field theory (George *et al.*, 1976; Vollhardt, 2012), show that the zero temperature Mott transition is not continuous, it is still worthwhile to investigate it from the perspective of the scaling theory. Different approximations to the Hubbard model (Hubbard, 1963, 1964a, 1964b), including Gutzwiller's theory (Gutzwiller, 1965), are examined from the point of view of the theory of quantum critical phenomena, which results in a better understanding of the nature of these approaches. The exact solution of the Hubbard model in one dimension combines with scaling to provide a new perspective to the problem of correlated electronic systems.

Superconductors can in general be driven to the normal state by pressure, magnetic field, impurities, etc. From this point of view they are of interest here and different kinds of superconducting quantum critical points will be studied.

Topological quantum phase transitions are radical examples of the failure of the Landau paradigm to describe these phase transitions. Specifically, the concept of an order parameter and of symmetry breaking does not apply for these transitions. However, we show that the renormalisation group remains a useful approach to this problem. The final chapters of the book deal with first-order quantum phase transitions. Although there is no diverging length and time in these transitions, we will show that scaling concepts continue to be applicable even if for a limited region of the phase diagram.

A fundamental theory that permeates and is the foundation of most of this book is the renormalisation group (RG). It is viewed here not only as a tool to calculate critical exponents, but as a framework that allows many features of quantum critical systems to be organised, unified and ultimately understood. The concepts of crossover, unstable fixed points, attractors, flow in parameter space and relevant or irrelevant fields turn out to be extremely useful to describe the physical behaviour of strongly correlated many-body systems, as we will show here.

The RG is particularly convenient in cases where the phase transition is not clearly associated with an order parameter, like in the case of metal–insulator transitions. Also, for topological quantum phase transitions that separate phases with the same broken symmetry but with trivial or non-trivial topological properties, the RG associates these transitions with an unstable zero-temperature fixed point and provides the theoretical apparatus necessary for their description. It turns out that the flow of the RG equations to the different attractors is sometimes sufficient to characterise the nature of the phases with no need to consider explicitly an order parameter. We start using the renormalisation group *language* from the beginning of the book. It is introduced as the material on phase transitions is being presented. In this way the reader is familiarised with the RG concepts before we get to the more technical chapters based on this technique.

The notion of *crossover* had an early application in the RG approach to the single impurity Kondo problem (Wilson, 1975). It gave a consistent explanation for the physics of the formation of local magnetic moments in metals and for the significance of the *Kondo temperature*. We will show that this concept remains extremely important also for the *Kondo lattice* model of heavy fermions in spite of the lattice translation invariance of this system. In this case it is associated with a new energy scale and with the idea of *coherence* that marks the onset of the Fermi liquid regime in dense Kondo systems. The existence of a magnetic quantum critical point and the scaling properties it engenders provide a simple interpretation of the concept of coherence in heavy fermions.

The scaling properties of a system close to a quantum phase transition can be derived considering a set of scale transformations and the existence of a characteristic length and a characteristic time that determine its behaviour near the quantum critical point. The renormalisation group provides the mathematical apparatus to describe these scale transformations. The critical point is associated with an *unstable fixed point* which governs the transition. The set of critical exponents associated with this fixed point characterises the *universality class* of the transition. Besides specifying the divergence of the correlation length, susceptibility, the critical slowing down, etc., for a many-body system, these exponents also characterise the critical behaviour of the compressibility, conductivity, superfluid density and the enhancement of the thermal mass close to the zero-temperature instability.

There is a large class of quantum transitions studied in this book for which an order parameter and its associated conjugate field cannot be clearly identified. This implies that not all the usual critical exponents can be defined. An example we will discuss is that of topological quantum phase transitions. We will show, however, that for this class of transitions, three exponents and the dimension d of the system suffice to characterise the quantum critical behaviour. The correlation length exponent ν, the dynamic exponent z and the critical exponent α that describes the behaviour of the singular part of the ground state energy density close to the QCP. These exponents are not independent but are related by the quantum hyperscaling relation $2 - \alpha = \nu(d + z)$ that plays an important role in this book.

In spite of an early application of scaling ideas in condensed matter, for example, in the problem of localisation due to disorder, it took some time for them to be used systematically in the study of strongly correlated electronic materials like heavy fermions and Mott insulators. Before we describe how these concepts can be useful to understand the physics of these many-body systems, it is instructive to derive the scaling theory of a zero-temperature phase transition considering a simple model borrowed from the theory of localised magnetism.

1.2 Renormalisation Group and Scaling Relations

Let us consider the simplest model which exhibits a quantum phase transition, namely the one-dimensional Ising model in a transverse magnetic field (Pfeuty, 1970):

$$H = -J \sum_i S_i^z S_{i+1}^z - h \sum_i S_i^x - H \sum_i S_i^z, \tag{1.1}$$

where $J > 0$ is the nearest neighbour coupling, h the transverse field and H the uniform magnetic field in the z-direction conjugate to the *order parameter* $< S^z >$. Let us consider initially the case $H = 0$. It is perfectly reasonable to expect, and in fact it does occur, that

- $T = 0$, $h = 0$. There is long-range magnetic order with an order parameter $< S^z > \neq 0$.

• $T = 0$, $h = \infty$. The transverse field destroys the long-range magnetic order and $< S^z >= 0$.

Then, one possibility, besides a smooth crossover, is that at a critical value of the ratio (h/J) there is a zero-temperature phase transition from an ordered state with $< S^z > \neq 0$ to a disordered state, i.e. to a state with a vanishing order parameter, $m =< S^z > = 0$. Here we are concerned with the latter case, although crossover effects will also be considered.

As we will see further on in this book, from the renormalisation group (RG) point of view, the phase transition at $(h/J) = (h/J)_c$ is connected with the existence of an *unstable zero-temperature fixed point*, at $(h/J)_c$, of the RG equations for the parameters of the model. Associated with this *quantum critical point* (QCP) we have a set of critical exponents, which describe the singular behaviour of the different physical quantities at the transition. We will find also from the renormalisation group transformations, *stable fixed points* at $(h/J) = 0$ and $(h/J) = \infty$, which are the *attractors* of the ordered and disordered phases respectively.

1.3 The Critical Exponents

In the vicinity of a second-order phase transition, either at finite or zero temperature, several physical quantities present non-analytical or diverging power law behaviour which are characterised by critical exponents (Ma, 1976). If g measures the distance to the critical point, the singular part of the free energy density f_s, the correlation length ξ, the critical relaxation time τ, the order parameter $m = -\partial f_s/\partial H$, the order parameter susceptibility $\chi = -\partial^2 f_s/\partial H^2$, where H is the field conjugated to the order parameter, behave, as $g \rightarrow 0$, in the following way (Stanley, 1971):

$$f_s \propto |g|^{2-\alpha}$$
$$\xi \propto |g|^{-\nu}$$
$$m \propto |g|^{\beta}$$
$$\chi \propto |g|^{-\gamma} \qquad (1.2)$$
$$\tau_\xi \propto |g|^{-\nu z}$$
$$m(H, g = 0) \propto H^{1/\delta},$$

where the last equation is defined at the critical point $|g| = 0$. These equations define the most common critical exponents, α, ν, β, δ and γ and the dynamic exponent z. There is another important exponent, the critical exponent η associated with the behaviour of the order parameter correlation function at the critical point.

Since at $g = 0$ there is no length scale, as the correlation length becomes infinite, the order parameter correlation function $G(r)$ decays algebraically with the distance r, in the following way, $G(r) \propto 1/r^{d-2+\eta}$. Alternatively, we can use the Fourier transform of $G(r)$, to define η through the relation $G(k) \propto k^{2-\eta}$. For the case of quantum transitions, the definition of η involves the exponent z, such that

$$G(r) \propto \frac{1}{r^{d+z-2+\eta}}$$

at the quantum critical point. In this quantum case we shall get used to seeing the combination $d + z$ playing the role of an effective dimensionality.

The critical exponents introduced above are not independent. They are in fact related by scaling relations, such as (Stanley, 1971; Ma, 1976)

$$\alpha + 2\beta + \gamma = 2$$
$$\beta + \gamma = \beta\delta$$
$$\nu(2 - \eta) = \gamma.$$

Particularly interesting is the hyperscaling relation which relates the critical exponents to the dimension d of the system. In the thermal case this is the Josephson relation:

$$2 - \alpha = \nu d. \tag{1.3}$$

For quantum phase transitions, as we show in this chapter, the hyperscaling relation is modified being given by (Continentino *et al.*, 1989)

$$2 - \alpha = \nu(d + z), \tag{1.4}$$

where again we find the combination $d + z$, playing the role of an effective dimensionality. Another important relation in the quantum case is (Fisher *et al.*, 1989)

$$2\beta = \nu(d + z - 2 + \eta), \tag{1.5}$$

which is also a hyperscaling relation since it involves the dimension of the system.

1.4 Scaling Properties Close to a Zero-Temperature Fixed Point

A fundamental step in the theory of critical phenomena is to understand the properties of physical systems under a change of scale in space and time dimensions. The reason is that, when the system approaches a critical point, the size of the regions in which the degrees of freedom are correlated in space and time increases. The typical size of these regions and their relaxation time define the correlation length ξ and the characteristic time τ_ξ. At the critical point all spins are correlated, the

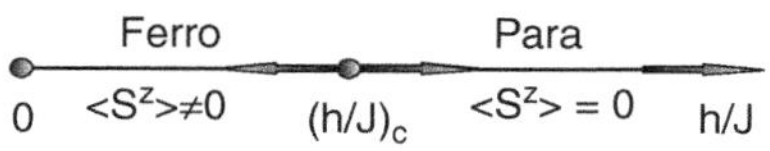

Figure 1.1 Zero-temperature phase diagram of the transverse field Ising model for $d \geq 1$. The phase transition at $(h/J)_c$ is associated with an unstable fixed point of the renormalisation group (RG) equations. The arrows show the direction of the flow of the RG equations. In the ordered ferromagnetic phase the flow is towards the strong coupling attractor at $J = \infty$. The flow in the paramagnetic phase is towards $h = \infty$ or $J = 0$.

correlation length diverges and the system is *scale invariant*. Then it turns out that for an understanding of critical phenomena, it is necessary to track the change of the physical parameters, like the interactions, magnetic field, magnetisation, etc. under a scale transformation (Fig. 1.1). Within the mathematical formalism of the RG, which describes the scale transformations, the scale-invariant quantum critical point is associated with a fixed point of the RG equations.

Consider the simple model above of Ising spins in a transverse field. As the system gets closer to the quantum critical point strongly interacting spins inside highly correlated regions can be treated as a single effective spin. In this case we can imagine a new lattice where the block of strongly interacting spins is replaced by a single effective spin. The question is, what is the new coupling between the effective spins in this new lattice? What is the magnetic field acting in these effective spins? Let us denote the physical parameters in this new effective lattice, with length scale L', by primed quantities. In order to relate the primed quantities to those on the original lattice, at length scale L, it is necessary to describe how they change under a redefinition of the length and time scale, since the primed and original lattices represent the same physical system at different length scales.

Let us consider how the parameters of the Hamiltonian 1.1 change, under a length scale transformation by a factor b, close to the zero-temperature phase transition. We have

$$\begin{cases} J' = b^{-y} J & \text{or } h' = b^{-y} h \\ g' = b^a g \\ H' = b^x H \end{cases} \tag{1.6}$$

where we have introduced three exponents y, a and x. The quantity $g = (h/J) - (h/J)_c$ measures the distance to the critical point in parameter space. In the equations of the first line, *which hold at the QCP*, we took into account that J and h scale with the same exponent, since $(h/J)' = (h/J)$, at the fixed point. The prime refers to the quantities renormalised under the length scale transformation. The way this transformation has been implemented, i.e. a physical quantity at a given length scale is related to the same quantity in another length scale, for example, $(H'/H) = (L/L')^x = b^x$, has been justified by Kadanoff *et al.* (1967).

The $T = 0$ free energy density (singular part) f_s can be written as

$$f_s = F_s/L^d = Jf(g, H/J),$$

where $f(q, p)$ is a scaling function and J is the coupling constant which has a dimension of energy. At zero temperature this is the available energy scale to fix the energy dimension. L is the size of the system and d its Euclidean dimension. The new free energy density and correlation length in the system where size has been rescaled by $L' = L/b$ are

$$f_s' = b^d f_s = J'f\left(|g|', H'/J'\right)$$
$$\xi'\left(|g|', H'/J'\right) = b^{-1}\xi\left(|g|, H/J\right).$$

Using the relations 1.6 in the equations above we get

$$\frac{f_s(|g|, H/J)}{J} = b^{-(d+y)}f\left[b^a|g|, b^{(x+y)}H/J\right]$$
$$\xi(g, H/J) = b\,\xi'\left[b^a|g|, b^{(x+y)}H/J\right].$$

Now since b is arbitrary, we take

$$b^a|g| = 1$$

or

$$b = |g|^{-1/a}$$

to obtain

$$\frac{f_s}{J} = |g|^{(d+y)/a}f\left[1, \frac{H/J}{|g|^{(x+y)/a}}\right] \tag{1.7}$$

and

$$\xi = |g|^{-1/a}\xi'\left[1, \frac{H/J}{|g|^{(x+y)/a}}\right]. \tag{1.8}$$

From our previous definition of the critical exponents, we can identify the correlation length exponent $\nu = 1/a$. From the definition of the exponent α associated with the singular part of the ground state energy density, i.e. $f_s/J = |g|^{2-\alpha}$, Eqs. 1.2 and since $1/a = \nu$, we get

$$2 - \alpha = \nu(d + y). \tag{1.9}$$

This is the quantum hyperscaling relation, which relates the critical exponents ν and α to the dimensionality of the system d and to the exponent y which renormalises the coupling J at the fixed point. It differs from the usual hyperscaling relation of finite temperature critical phenomena in a very fundamental way, since d is replaced by $d + y$. We shall return to this important point later.

The order parameter, in the present case the magnetisation, $m = < S^z >$, is defined by the derivative of the free energy with respect to its conjugate field H

$$m = -(\partial f_s / \partial H)_{H=0} \propto |g|^\beta.$$

Taking the derivative of Eq. 1.7 we get

$$m = -(\partial f_s / \partial H)_{H=0} = |g|^{(d+y)/a} |g|^{-(x+y)/a} f'(1, 0) \propto |g|^{\nu(d-x)},$$

where $f'(1, 0) = (\partial f(1, p)/\partial p)_{p=0}$; consequently

$$\beta = \nu(d - x). \tag{1.10}$$

In the presence of the external field, we have

$$m(H) = -(\partial f_s / \partial H) = |g|^{\nu(d-x)} f' \left[1, \frac{H/J}{|g|^{(x+y)/a}} \right].$$

At the critical point, $g = 0$, the order parameter should be finite in the presence of the external conjugate field. We may then write

$$m(H) = |g|^{\nu(d-x)} \left(\frac{H/J}{|g|^{(x+y)/a}} \right)^r,$$

where the exponent r is obtained from the condition that the dependence on $|g|$ cancels out to yield a finite order parameter at $|g| = 0$, i.e.

$$\frac{d - x}{a} - r \frac{x + y}{a} = 0$$

since $\nu = 1/a$. We then find

$$r = \frac{d - x}{x + y} = \frac{\beta}{\Delta}, \tag{1.11}$$

where we introduced the exponent $\Delta = \nu(x + y)$ and used Eq. 1.10. Since the critical exponent δ is defined through the dependence of the order parameter on the conjugate field at the critical point, i.e. $m(H, |g| = 0) \propto H^{1/\delta}$, we get, using $\delta = 1/r$, the scaling relation, $\delta = \Delta/\beta$.

The susceptibility is given by

$$\chi = - \left(\frac{\partial^2 f_s}{\partial H^2} \right)_{H=0} \propto |g|^{-\gamma}.$$

Taking the second derivative of Eq. 1.7, we find

$$\gamma = \nu(2x + y - d). \tag{1.12}$$

Equations 1.9, 1.10, 1.11 and 1.12 yield the standard scaling relation

$$\alpha + 2\beta + \gamma = 2, \tag{1.13}$$

and the scaling law

$$\Delta = \beta + \gamma. \tag{1.14}$$

Irrelevant Variables

Let us consider an interaction or parameter of a system that is close to the critical point scales as

$$u' = b^{-s}u \tag{1.15}$$

under a change of length scale by a factor b. In the scaling form, for example, of the correlation length this interaction will appear as

$$\xi = |g|^{-1/a}\xi'\left[1, \frac{H/J}{|g|^{(x+y)/a}}, u|g|^{s/a}\right], \tag{1.16}$$

as can be easily verified from Eq. 1.8. This can still be written as

$$\xi = |g|^{-\nu}\xi'\left[1, \frac{H/J}{|g|^{\beta+\gamma}}, u|g|^{\nu s}\right], \tag{1.17}$$

where we used, $\nu = 1/a$ and $\Delta = \nu(x+y) = \beta + \gamma$. Notice that as the critical point is approached, i.e. $|g| \to 0$, the scaling variable $u|g|^{\nu s}$ goes to zero and the interaction u becomes *irrelevant* at the critical point, provided the scaling function $\xi'(q, p, v \to 0)$ is well behaved in this limit. In this case the *irrelevant interaction* u will give rise essentially to *scaling corrections*. However, in the situation where $\xi'(1, p, v \to 0)$ is not well behaved, for example, $\xi'(1, p, v \to 0) \propto 1/v$, we cannot neglect the effect of the interaction u close to criticality. In this case, we refer to u as a *dangerously irrelevant interaction* and we anticipate that we will find this situation at different points in this book.

The Special Role of Time and the Dynamic Exponent

Close to the zero-temperature critical point, time τ scales as

$$\tau' = b^{z}\tau,$$

which defines the dynamic exponent z. The quantum character of the critical fluctuations allows us to relate z to the exponent y governing the scaling of the coupling constant J at the quantum critical point, as we now show.

Since the scale for the energy fluctuations at $T = 0$ is set by the coupling J, we expect

$$\Delta E' = b^{-y} \Delta E$$

while

$$\Delta \tau' = b^z \Delta \tau.$$

Now we want the uncertainty relation $\Delta E \Delta \tau \geq \hbar$ to be a scaling invariant, i.e.

$$\Delta E' \Delta \tau' = b^{(z-y)} \Delta E \Delta \tau \geq \hbar$$

and consequently we must have:

$$y = z$$

and the quantum hyperscaling relation, Eq. 1.9, can then be written as

$$2 - \alpha = \nu(d + z). \qquad (1.18)$$

In this relation the dimension d is replaced by $d_{eff} = d + z$, which plays the role of an *effective dimensionality*. This shift in the dimensionality has important consequences for quantum phase transitions:

- It implies that the exponents of the quantum system are the same of the corresponding classical one in $d_{eff} = d + z$ dimensions (Hertz, 1976). For example, for the $d = 1$ Ising model in a transverse field, the critical exponents associated with the zero-temperature unstable fixed point take the values, $\beta = 1/8$, $\alpha = 0$, $\nu = 1$, $\gamma = 1.75$ which can identified as the exact critical exponents obtained by Onsager for the classical Ising model in two dimensions. So in this case we expect to find $z = 1$, which is indeed the case (Pfeuty, 1970; Young, 1975).
- Since d_{eff} is increased, it may reach the *upper critical dimension*, d_c, above which fluctuations can be ignored and mean field gives an appropriate description of the system. In this case the exponents associated with the $T = 0$ fixed point assume mean field or Gaussian values (Hertz, 1976). This is actually what happens for some important phase transitions considered here, as we will discuss further on.

The Correlation Function at $T = 0$

The fluctuation-dissipation theorem gives us the following relation between the wave vector-dependent static susceptibility $\chi(q)$ and the dynamic q-dependent order parameter correlation function $S(q, \omega)$ at $T = 0$ (Ma, 1976):

$$\chi(q) = \int \frac{d\omega}{2\pi} \frac{S(q, \omega)}{\omega}. \qquad (1.19)$$

The singularity of $\chi(q = 0) \propto |g|^{-\gamma}$ (for a ferromagnet) defines the exponent γ. On the other hand, the static correlation function is given by

$$S(q) = \int d\omega S(q, \omega) \qquad (1.20)$$

and

$$S(q = 0) \propto |g|^{-\gamma_s},$$

which defines γ_s. For finite temperatures, we have $\chi T = S(q = 0)$ and $\gamma = \gamma_s$. This is not the case, however, at $T = 0$. Let us consider the scaling *ansatz* for $S(q, \omega)$ which defines the exponent η:

$$S(q, \omega) = \xi^{2-\eta} D\left(q\xi, \omega\xi^z\right). \qquad (1.21)$$

Taking this scaling expression in Eqs. 1.19 and 1.20, we get the relations

$$\gamma_s = (2 - z - \eta)\nu = \gamma - \nu z \qquad (1.22)$$

and

$$2\beta = \nu(d + z - 2 + \eta) \qquad (1.23)$$

at zero temperature. Note that for $T \neq 0$ we have $\gamma_s = \gamma = (2 - \eta)\nu$. Eq. 1.22 implies $G(r) = \frac{1}{r^{d+z-2+\eta}} g(r/\xi)$ since, $S(q = 0) \equiv \int dr\, G(r)$. Comparing with the finite temperature case we see again that $d + z$ plays the role of an effective dimension for quantum transitions.

1.5 Extension to Finite Temperatures

We would like to extend the scaling approach for small but finite temperatures. Since temperature is a parameter, it is renormalised by the characteristic energy or coupling constant at the zero-temperature fixed point. We then have

$$\left(\frac{T}{J}\right)' = b^z \left(\frac{T}{J}\right). \qquad (1.24)$$

Formally, we may think of the renormalised temperature as a *field* which renormalises under a scale transformation according to Eq. 1.24. It is interesting here to consider three possibilities depending on the value assumed by the exponent y:

- $z > 0$. In this case the flow of the renormalisation group equations is away from the zero-temperature fixed point. In the renormalisation group language we say temperature is a *relevant field*. This is the case for the Ising model in a transverse field ($y = z = 1$) and also for the many-body problems we are concerned with here.

- $z = 0$. This is the marginal case generally associated with the collapse of a finite temperature fixed point and the one at zero temperature. It is characterised by the appearance of logarithmic corrections. Such a situation occurs, for example, at the isotropic fixed point of the anisotropic Heisenberg ferromagnet in two dimensions (de Mello and Continentino, 1990).
- $z < 0$. This is a peculiar situation which may occur in random classical magnetic systems. It leads to dimensional reduction ($d_{eff} < d$) and gives rise to anomalous critical slowing down for the finite temperature transition controlled by the $T = 0$ fixed point. This is the case of the Ising ferromagnet in a random field for $d > 2$ (Bray and Moore, 1985).

Let us now obtain how temperature will appear in the scaling functions. Returning to Eqs. 1.7 and 1.8 with the additional renormalisation equation for temperature, Eq. 1.24, we get, for example, for the temperature dependent correlation length:

$$\xi = |g|^{-\nu} f_\xi \left[\frac{T/J}{|g|^{\nu z}}, \frac{H/J}{|g|^{\Delta}} \right], \tag{1.25}$$

where $f_\xi(q, p)$ is a scaling function.

For the free energy density, we find

$$f = |g|^{2-\alpha} f_E \left[\frac{T/J}{|g|^{\nu z}}, \frac{H/J}{|g|^{\Delta}} \right], \tag{1.26}$$

where $\Delta = \beta + \gamma$ and we used $y = z$. We will find it convenient to define a crossover exponent $\phi = \nu y = \nu z$.

Note in the equation above a very interesting feature, which is the appearance of the dynamic critical exponent z in a static thermodynamic quantity, namely the free energy. This is the signature of the quantum phase transition and of the inextricability of static and dynamics for these transitions.

The Crossover Line and the Exponent $\phi = \nu z$

Let us consider the two-dimensional Ising model in a transverse field. As will be shown in Chapter 3, it is possible to make an expansion of the RG equations for the parameters of the model close to the zero-temperature unstable fixed point K_c associated with the quantum phase transition. This expansion, quite generally, takes the simple form of a recursion relation:

$$K_{n+1} = b^a (K_n - K_c) + K_c,$$

which starts to be iterated from an arbitrary point $K_0 = (h/J)$, sufficiently close to the critical or unstable fixed point $K_c = (h/J)_c$ for the expansion to hold. Physically this equation can be seen as describing the change of the ratio between

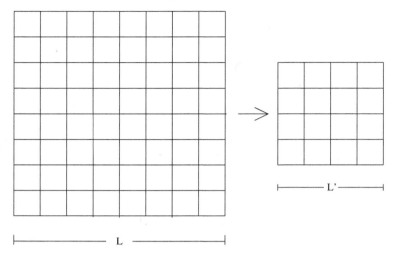

Figure 1.2 A scaling transformation, by a factor b, which takes a lattice of size L to size L'. In the present case $b = 2$. The figure represents the same physical system where the unit of the length scale has been changed by the factor $b = L/L' = 2$.

the transverse field and the coupling constant, close to the critical point, as the length scale of the system changes by a factor b (see Eqs. 1.6 and Fig. 1.2). Notice that for $K_0 = K_c$, we get $K_1, K_2, \cdots, K_N = K_c$. Then the critical point K_c is the *fixed point* of the recursion relation and does not renormalise under the length scale transformation. This is the mathematical expression of the scale invariance of the critical point, as it manifests in the RG approach. The RG associates a mathematical entity, the unstable fixed point of a recursion relation, with a physical quantity, namely, the critical point of a physical system.

For finite but small temperatures, we can generalise the equation above. Expanding to lowest order in (T/J), we get

$$K_{n+1} = K_c + b^a(K - K_c) - \widetilde{T}_n^2 \tag{1.27}$$
$$\widetilde{T}_{n+1} = b^y \widetilde{T}_n, \tag{1.28}$$

where $\widetilde{T} = (T/J)$. Notice that the first term in the expansion for small temperatures in the equation for K_{n+1} enters as an analytic T^2 contribution. In fact, the purpose of the renormalisation group is to produce non-analytic behaviour starting from analytic expansions. The equations above can be iterated in the following way:

$$K_1 = K_c + b^a(K_0 - K_c) - \widetilde{T}_0^2$$
$$\widetilde{T}_1 = b^y \widetilde{T}_0$$

and

$$K_2 = K_c + b^a(K_1 - K_c) - \widetilde{T}_1^2$$
$$\widetilde{T}_2 = b^y \widetilde{T}_1.$$

Using Eq. (1.20), the equation for K_2 can be rewritten as

$$K_2 = K_c + b^a[b^a(K_0 - K_c) - \widetilde{T}_0^2] - b^{2y}\widetilde{T}_0^2$$
$$K_2 = K_c + b^{2a}(K_0 - K_c) - b^a\widetilde{T}_0^2 - b^{2y}\widetilde{T}_0$$
$$K_2 = K_c + b^{2a}(K_0 - K_c) - \frac{(b^a + b^{2y})(b^a - b^{2y})}{(b^a - b^{2y})}\widetilde{T}_0^2$$
$$K_2 = K_c + b^{2a}(K_0 - K_c - \frac{1}{b^a - b^{2y}}\widetilde{T}_0^2) + \frac{1}{b^a - b^{2y}}b^{4y}\widetilde{T}_0^2.$$

Repeating the iteration n times we get

$$K_n = K_c + b^{na}(K_0 - K_c - a_0\widetilde{T}_0^2) + a_0(b^{ny}\widetilde{T}_0)^2,$$

where $a_0 = 1/(b^a - b^{2y})$, $K_0 = h/J$ and $\widetilde{T}_0 = T/J$ (we assume that $y/a = vy = vz > 1/2$, such that $a_0 > 0$). Taking $\ell = b^n$ we finally obtain:

$$K_\ell = K_c + \ell^a(K_0 - K_c - a_0\widetilde{T}_0^2) + a_0(\ell^y\widetilde{T}_0)^2. \qquad (1.29)$$

Since ℓ is arbitrary, we repeat the scaling procedure until

$$\ell^a(K_0 - K_c - a_0\widetilde{T}_0^2) = 1$$

and this length scale defines the correlation length

$$\ell = \xi = \frac{1}{(K_0 - K_c - a_0\widetilde{T}_0^2)^{1/a}} \propto [h - h_c(\widetilde{T})]^{-\nu}$$

with the correlation length exponent $\nu = 1/a$ and $h_c(T) = h_c + a_0T^2$. Substituting the expression for ξ into Eq. 1.29, we find that at length scale ξ,

$$K_\xi = K_c + 1 + a_0\left(\frac{\widetilde{T}}{[h - h_c(T)]^{\nu y}}\right)^2. \qquad (1.30)$$

Let us consider the *non-critical region* of the phase diagram, for $(h/J) > (h/J)_c$, where no physical quantity diverges. Notice that if $\frac{\widetilde{T}}{[h-h_c]^{\nu y}} << 1$, K_ξ is essentially a constant independent of temperature. On the other for $\frac{\widetilde{T}}{(h-h_c)^{\nu y}} >> 1$, K_ξ acquires a temperature dependence. The same holds for any scaling function $f(K_\xi)$ whenever K_ξ appears as an argument. Consequently the line

$$T_{cross} = (h - h_c)^\phi, \qquad (1.31)$$

where $\phi = vy = vz$ represents a crossover line which separates two different regimes for the behaviour of K_ξ or for any scaling function $f(K_\xi)$, in the non-critical part of the phase diagram, i.e. for $(h/J) > (h/J)_c$.

The existence of a crossover line in the non-critical region of the phase diagram, associated with a quantum critical point, is a general feature of quantum phase transitions. What is the mathematical significance of this crossover line, since there is no phase transition occurring along it? Consider a physical quantity X written in the scale invariant form as $X = | g |^{-x} f(T/T_{cross})$ where $T_{cross} = |g|^{vz}$ and $g = 0$ defines the critical point. Taking the derivative with respect to temperature yields $\partial X / \partial T = | g |^{-x-vz} f'(t)$ where $f'(t)$ is the derivative of the scaling function $f(t)$ with respect to $t = T/T_{cross}$. If we equate $\partial X / \partial T$ to zero to find the temperature T_m of the extrema of the quantity X, we get $f'(t_m) = 0$, besides trivial roots. This equation will have a solution, let's say for $t_m = C$, where C is a constant. This yields $T_m = C T_{cross} = C|g|^{vz}$ implying that the extrema of the quantity X occur along the crossover line. The same holds for inflection points. Consequently any anomaly on physical quantities, like maxima, for example, in the non-critical region of the phase diagram will occur along the crossover line making this line accessible experimentally. Although the constant C may depend on the particular physical quantity, as expected for a crossover effect, the relevant, universal information is contained in the crossover exponent, $\phi = vz$, which is determined by the universality class of the transition. As for the physical significance of the crossover line, this will depend on the specific nature of the critical system.

The Critical Line and the Shift Exponent ψ

From the expression for the finite temperature correlation length obtained above, namely

$$\xi = \frac{1}{(K_0 - K_c - a_0 T_0^2)^v}$$

with $v = 1/a$, we obtain the equation for the *critical line*, T_c, in the $T \times g$ phase diagram, as the set of temperatures at which the correlation length diverges. This is given by $K_0 - K_c - a_0 T_c^2 \equiv g(T_c) = 0$, or $(h/J) = (h/J)_c - |a_0|T_c^2$ or still $T_c \propto |(h/J) - (h/J)_c|^{1/2} = |g(T = 0)|^{1/2}$. The critical line in this case is analytic in temperature, with a T^2 temperature dependence. As we will see further on, this is not usually the case. Also, the exponent $\tilde{v}$, which describes the divergence of the correlation length at finite temperatures, is different from $v = 1/a$ associated with the QCP as obtained in the simple treatment above. In general, the critical line is given by, $T_c = |g|^\psi$, which defines the *shift exponent* ψ. In the case treated above, $\psi = 1/2$. In Fig. 1.3, we show a typical phase diagram in the $T \times g$ plane for the

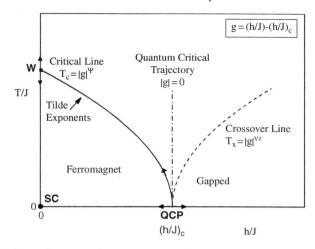

Figure 1.3 Phase diagram of the transverse field Ising model for $d > d_L = 1$. The arrows indicate the flow of the renormalisation group equations. Note that the zero-temperature fixed point is fully unstable. The semi-stable fixed point W governs the finite temperature transitions at the critical line $T_c = |g|^\psi$. The *tilde* exponents are associated with this fixed point and belong to a different universality class of that associated with the quantum critical point (QCP) at $(h/J)_c$, $T = 0$. SC is the strong coupling, attractive fixed point of the ferromagnetic phase.

Ising model in a transverse field for $d > 1$. The arrows illustrate the direction of the flow of the renormalisation group equations. Notice that this flow is always away from the quantum critical point at $g \equiv g(T = 0) = 0$ showing the fully unstable nature of the fixed point associated with the quantum critical point (QCP) at $g = 0$, $T = 0$. There is another very special feature to be noted from the direction of these flows. This has to do with the flow along the critical line of finite temperature phase transitions, $T_c(g)$. The fact that the flow along this line is away from the QCP implies that the critical behaviour along this line is governed by another fixed point which controls the finite temperature critical behaviour. Consequently the thermal critical exponents, associated with the finite temperature phase transitions are in general different from those associated with the QCP, i.e. they belong to a different universality class.

There are situations where the critical line is governed by the same exponent of the crossover line, i.e. $\psi = \nu z$, as in the two-dimensional Ising model in a transverse field. This equality is the content of the so-called *generalised scaling hypothesis*, which, however, does not hold for most of the problems discussed in this book.

When dealing with the extension of quantum phase diagrams to finite temperatures we have to be aware of the *lower critical dimension d_L* of the system under study. This is the dimension at and below which there is no finite temperature transition in the system as the zero-temperature ordered phase is immediately destroyed by thermally excited low-energy modes ($d_L = 1$ for the transverse field Ising model and $d_L = 2$ for the Heisenberg model). In this case there is no critical line at finite T and the behaviour of the system as temperature is reduced to 0 is considerably modified. Another important concept that we mentioned before is that of *upper critical dimension*, d_c, above which the exponents become mean field-like. In general, this dimension coincides with that for which the hyperscaling relation is satisfied using the mean field values of the critical exponents for the correlation length and free energy, ν and α, respectively, for a given type of phase transition. Since the hyperscaling relations for thermal and quantum phase transitions are different, they will have different upper critical dimensions.

1.6 Temperature-Dependent Behaviour near a Quantum Critical Point

Naive Scaling

An interesting feature of the scaling approach is that it allows the singular behaviour of the physical quantities of interest to be determined as a function of temperature at criticality. Let us consider the uniform susceptibility of the Ising model in a transverse field along the *quantum critical trajectory*, i.e. as it approaches the critical point at $(h/J)_c$ along the line $|g| = 0$, $T \to 0$. The general scaling form for the susceptibility is $\chi = |g|^{-\gamma} f(T/T_{cross})$ with $T_{cross} \propto |g|^{\nu z}$. In order to have a non-trivial result for χ along the trajectory, $|g| = 0$, $T \to 0$, we require that the dependence of χ on $|g|$ cancels out. For this purpose, the scaling function $f(T/T_{cross}$ is expanded as $f(T/T_{cross}) \approx (T/T_{cross})^x$ such that $\chi = |g|^{-\gamma}[T/T_{cross}]^x = |g|^{-\gamma}[T/|g|^{\nu z}]^x$ or $\chi = |g|^{-(\gamma+\nu z x)}T^x$. Finally from the condition that the dependence on $|g|$ cancels out, we determine the exponent x. This condition is, $\gamma + \nu z x = 0$, which yields $x = -\gamma/\nu z$, implying that the susceptibility diverges as $\chi = T^{-\gamma/\nu z}$, at the critical point $|g| = 0$, with decreasing temperature.

1.7 Generalised Scaling

The results above are not completely general. They rely on some assumptions that can be made clear when extending a general scaling theory (Pfeuty, Jasnow and Fisher, 1974) for a quantum critical point including finite temperatures (Continentino, 1990). Let us consider a system above its lower critical dimension $d > d_L$

with a line of finite temperature phase transitions in its phase diagram. For concreteness imagine the ferromagnetic quantum transverse Ising model at dimension $d = 2$ with a phase diagram, as shown in Fig. 1.3. There is a quantum critical point at $T = 0$, $g = 0$ and a critical line $T_c(g)$ of finite temperature phase transitions separating the paramagnetic and ferromagnetic phases. Close to the QCP this line is governed by the equation, $T_c = (1/u)|g|^\psi$, where u is a constant and ψ is the shift exponent. In many cases of interest, as we show in Chapter 5, u is the coefficient of a quartic interaction in the expansion of the action or free energy in terms of an effective order parameter and that turns out to be *dangerously irrelevant*. There are three fixed points in the phase diagram shown in Fig. 1.3 (dos Santos, 1982). The fully unstable QCP at $T = 0$, $g = 0$, a semi-stable fixed point W at finite temperatures (T_c/J), $(h/J) = 0$ and a fully stable strong coupling fixed point SC at $T = 0$, $(h/J) = 0$ that is the attractor of the ferromagnetic phase. Consider the zero-temperature ferromagnetic transition at $(h/J)_c$. For small temperatures T, and sufficiently close to the QCP, the critical line can be written as $(h/J) = (h/J)_c - uT^{1/\psi}$, or alternatively as $g(T_c) = 0$, where $g(T) = (h/J) - (h/J)_c + uT^{1/\psi}$. In Fig. 1.3 the arrows indicate the flow of the RG equations. At finite temperatures along the critical line the RG flow is away from the QCP, towards the semi-stable Ising fixed point W. In the language of the RG, temperature is a *relevant field* at the QCP since it renormalises away from it under a RG transformation. At the semi-stable fixed point W, the rate of flow away from it determines the thermal critical exponents which control the critical behaviour all along the critical line of finite temperature transitions $T_c(g)$. These thermal exponents are generally different from those associated with the QCP. They will be denoted by a *tilde* to distinguish them from the critical exponents associated with the QCP. We will thus refer to the quantum exponents as $\{\alpha, \beta, \gamma, \nu, z, \cdots\}$ and to the thermal critical exponents as $\{\tilde{\alpha}, \tilde{\beta}, \tilde{\gamma}, \tilde{\nu}, \tilde{z}, \cdots\}$.

In the first set we emphasise the special role of the dynamic quantum critical exponent z, which appears in the expression for the free energy, Eq. (7.4). Its classical (thermal) counterpart $\tilde{z}$ has no special thermodynamic significance. It does not affect the critical behaviour of static thermodynamic quantities and its unique effect is to determine the critical slowing down close to the thermal transition. The fundamental character of the dynamic exponent in quantum phase transitions is a consequence of the inextricability of static and dynamics in quantum critical phenomena expressed, for example, in the uncertainty relation $\Delta E \Delta \tau \geq \hbar$.

In each set, the critical exponents are bound by scaling relations, for example, $\alpha + 2\beta + \gamma = 2$ and $\tilde{\alpha} + 2\tilde{\beta} + \tilde{\gamma} = 2$, which reduce the number of independent critical exponents. Particularly important is the hyperscaling relation involving the dimensionality d of the system. In the classical case the hyperscaling relation is $2 - \tilde{\alpha} = \tilde{v}d$. The quantum hyperscaling relation in turn involves the dynamic quantum critical exponent and is given by $2 - \alpha = v(d + z)$ (Continentino *et al.*, 1989) as we have already shown.

The above discussion appears to suggest that classical and quantum phase transitions in general have separate descriptions not interfering with each other. That this is not the case will be shown below, on the basis of a generalised scaling approach (GSA) (Pfeuty, Jasnow and Fisher, 1974) extended for quantum phase transitions (Continentino, 1990). More specifically, we will see that at low temperatures, in the neighbourhood of the QCP, if the system is above its lower critical dimension d_L and presents a line of second order thermal phase transitions, the description of the quantum and thermal critical behaviour may involve, in each case, both quantum and thermal critical exponents.

In general this occurs if the effective dimension of the quantum phase transition is above its upper critical dimension, i.e. if $d + z \geq d_c$. In this case, the generalised scaling relation $\psi = vz$ is violated due to a dangerously irrelevant interaction (Millis, 1993) requiring, as we show below, both quantum and thermal exponents to describe the physical behaviour along the quantum critical trajectory, $g = 0$, $T \to 0$.

For the transverse field ferromagnetic Ising model, the longitudinal uniform susceptibility $\chi(g, T)$ plays the role of the order parameter susceptibility. Close to the QCP at $(h/J)_c$ the generalised scaling form of this susceptibility is given by (Continentino, 2011)

$$\chi(g, T) \propto |g(T)|^{-\gamma} Q\left(\frac{T}{|g(T)|^{vz}}\right), \tag{1.32}$$

where $g(T) = |(h/J) - (h/J)_c + uT^{1/\psi}|$ is the normalised distance to the critical line $g(T_C) = 0$ and $g = g(T = 0)$.

Continuity imposes the following asymptotic behaviours of the scaling function $Q(t)$ $(t = T/|g(T)|^{vz})$:

- $Q(t = 0) = $ constant. This guarantees that at zero temperature we obtain the correct expression for the quantum critical behaviour of the longitudinal uniform susceptibility, $\chi(g, 0) \propto |g|^{-\gamma}$, such that, the order parameter susceptibility diverges at zero temperature with the quantum critical exponent γ.
- $Q(t \to \infty) \propto t^y$. The exponent $y = (\tilde{\gamma} - \gamma)/vz$ yields the correct thermal critical behaviour of the susceptibility close to the critical line $g(T) = 0$, i.e. $\chi(g, T) \propto A(T)|g(T)|^{-\tilde{\gamma}}$ as $T \to T_c(h)$.

The amplitude $A(T) \propto T^{(\tilde{\gamma}-\gamma)/\nu z}$ is determined by both classical (*tilde*) and quantum critical exponents. Then, close to the critical line, the order parameter susceptibility diverges as

$$\chi \propto A(T)|g(T)|^{-\tilde{\gamma}} \tag{1.33}$$

with the correct thermal critical exponent $\tilde{\gamma}$.

We now consider the quantum critical trajectory in the phase diagram of Fig. 1.3. We let the system to sit at the QCP, i.e. we fix (h/J) at $(h/J)_c$ and lower the temperature. Along this trajectory $g(0) = 0$ or $(h/J) = (h/J)_c$, and $T \to 0$, using the expression for the amplitude $A(T)$, we find from Eq. 1.33 that the susceptibility diverges as

$$\chi(0, T) \propto T^{\frac{\tilde{\gamma}-\gamma}{\nu z}} (uT^{\frac{1}{\psi}})^{-\tilde{\gamma}}. \tag{1.34}$$

Then, it follows that for $d + z > d_c$, in the presence of a dangerously irrelevant interaction u that breaks down the generalised scaling relation, such that $\psi \neq \nu z$, both thermal and quantum critical exponents are required to give the correct critical behaviour of a physical quantity along the QCT. Notice that for $\psi = \nu z$, the exponent $\tilde{\gamma}$ cancels out in the expression above and the divergence of the order parameter susceptibility, $\chi(0, T) \propto T^{-\gamma/\nu z}$ is governed only by the critical exponents associated with the QCP. The temperature dependence of this result coincides with that of *naive* scaling or of a *purely Gaussian* theory, in which the coefficient of the quartic interaction $u = 0$. The equality between the crossover and the shift exponents $\nu z = \psi$, i.e. the generalised scaling relation is expected to hold for $d + z < d_c$.

A similar analysis yields, for the correlation length close to the critical line,

$$\xi \propto A_L(T)|g(T)|^{-\tilde{\nu}} \tag{1.35}$$

with $A_L(T) = T^{(\tilde{\nu}-\nu)/\nu z}$. Along the quantum critical trajectory,

$$\xi(0, T) \propto T^{\frac{\tilde{\nu}-\nu}{\nu z}} (uT^{\frac{1}{\psi}})^{-\tilde{\nu}}, \tag{1.36}$$

we find the same interference between classical and quantum critical behaviour for $\psi \neq \nu z$. Again, for $\psi = \nu z$ the temperature dependence of $\xi(0, T)$ reduces to that of the purely Gaussian case, $\xi(0, T) \propto T^{-1/z}$. Notice that in Eqs. 1.34 and 1.36 u appears in the denominator, justifying its dangerously irrelevant character. For an application of generalised scaling to a quantum Bose glass to superfluid transition, see Yu *et al.* (2012).

A well known result for the correlation length (Millis, 1993) can be obtained from Eq. 1.36 assuming that the thermal and quantum correlation length exponents are Gaussian, i.e. $\tilde{\nu} = \nu = 1/2$. This gives

$$\xi(0, T)^{-2} = uT^{1/\psi}.$$

The same assumption for the susceptibility, i.e. taking $\tilde{\gamma} = \gamma = 1$ in Eq. 1.34 yields

$$\chi(0, T)^{-1} = uT^{1/\psi}.$$

Eqs. 1.33 and 1.35 and the respective expressions for the amplitudes give rise to a *quantum suppression of classical fluctuations*. Indeed, if as expected the *tilde* exponents are larger than the zero-temperature exponents, the amplitudes always decrease as we lower the temperature, suppressing the thermal singularities along the critical line as the QCP is approached.

The specific heat requires a more careful analysis since it vanishes in the zero-temperature axis. Its scaling behaviour is obtained from the scaling expression for the singular part of the free energy close to the critical line which is given by

$$f_S \propto T^{\frac{\tilde{\alpha}-\alpha}{\nu z}} |g(T)|^{2-\tilde{\alpha}}. \tag{1.37}$$

The most singular contribution for the specific heat close to this line is given by

$$C/T \propto T^{\frac{\tilde{\alpha}-\alpha}{\nu z}} |g(T)|^{-\tilde{\alpha}} \left(g'(T)\right)^2, \tag{1.38}$$

where $g'(T) = \partial g/\partial T = -(u/\psi)T^{(1/\psi)-1}$. In the Gaussian approximation for both thermal and quantum transitions, using $\tilde{\alpha} - \alpha = \nu z$ derived from hyperscaling, we obtain that close to the critical line,

$$C/T \propto T|g(T)|^{-\tilde{\alpha}} \left(g'(T)\right)^2. \tag{1.39}$$

Along the QCT, if $\psi = \nu z$, the temperature dependence of this contribution coincides with that of the *naive* approach or purely Gaussian result, namely $C/T \propto T^{(d-z)/z}$.

The simplest case of quantum suppression of classical fluctuations is the decrease of the mean-field jump in the specific heat along the critical line as the system approaches a quantum critical point. As an example, consider a superconductor that approaches a superconductor quantum critical point (SQCP) driven by external pressure P. From Eq. 1.38 for the specific heat, with $\tilde{\alpha} = \alpha = 0$, the critical line given by $T_c(P) \propto \sqrt{|P_c - P|}$, such that, $\psi = 1/2$, we find that along this line, $T_c(P)$, the specific heat jump $\Delta C_P \propto T_c(P)^3$. Normalising by the jump at $P = 0$ one finds the relation

$$\frac{\Delta C_P}{\Delta C_{P=0}} = \left(\frac{T_c(P)}{T_c(P = 0)}\right)^3 \tag{1.40}$$

such that the jump is reduced as the SQCP is approached, as can be seen in Fig. 1.4.

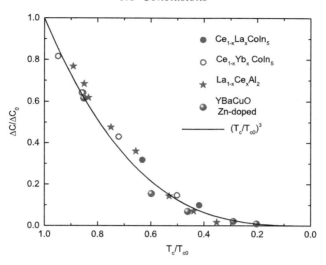

Figure 1.4 Normalised specific heat jump versus normalised temperature of some superconductors as T_c is reduced (see Eq. 1.40) Data from Refs. (Bauer *et al.*, 2011; Loram *et al.*, 1990; Luengo *et al.*, 1972).

1.8 Conclusions

We have presented in this chapter a scaling theory of quantum critical phenomena. We have shown that, close to a zero-temperature phase transition, the thermodynamic and correlation functions are governed by critical exponents associated with a quantum critical point. These exponents obey scaling relations and most remarkable is the appearance of the dynamic critical exponent in the hyperscaling relation, Eq. 1.18. This is due to the entanglement between static and dynamic properties which ultimately can be attributed, as was shown, to the Heisenberg uncertainty relation. The fundamental implication of this entanglement is the appearance of the dynamic exponent in the scaling form of the free energy, Eq. 1.26, from which all measurable thermodynamic properties can be derived.

As shown in this chapter, for systems near a QCP and for which $\psi \neq \nu z$, the temperature dependence of physical quantities along the quantum critical trajectory (QCT), $g = 0$, $T \to 0$, is determined by both, the *tilde* or thermal exponents and those associated with the QCP. Only when the generalised scaling hypothesis $\psi = \nu z$ holds is the behaviour along the QCT determined exclusively by the critical exponents of the zero-temperature phase transition. The breakdown of the relation $\psi = \nu z$ occurs in general when the effective dimension $d_{eff} = d + z > d_c$, where d_c is the *upper critical dimension*. This is commonly due to a dangerously irrelevant interaction, as will be shown in Chapter 5, and in this case the general scaling relations studied above should be considered.

For systems with a line of finite temperature phase transitions, we have shown that there is a *quantum suppression of classical fluctuations* close to the critical line. This is a direct consequence of the inequality between thermal and quantum critical exponents which are associated with different unstable fixed points.

In the next chapters we use the ideas introduced above to study quantum phase transitions in strongly correlated electronic materials. It turns out that when these systems are metallic in the non-critical side of their phase diagram, the crossover temperature can be identified with a *coherence temperature* below which they present Fermi liquid (FL) behaviour. This Fermi liquid however has its parameters strongly renormalised due to the proximity of the quantum critical point. We use scaling arguments to obtain how the exponents which govern the renormalisation of these parameters are related to the usual critical exponents associated with a zero-temperature phase transition as introduced above. As expected the dynamic exponent z also plays a special role in these many-body systems and shows up explicitly in the thermodynamic quantities. We shall study the Kondo lattice problem, which is a useful model to describe heavy fermion materials. We will also discuss a scaling theory for the metal–insulator transition due to correlations (Mott transition), giving special attention to charge fluctuation effects.

2
Landau and Gaussian Theories

2.1 Introduction

We have seen that in the vicinity of a second-order phase transition, either at finite or zero temperature, several physical quantities exhibit non-analytical or diverging power law behaviour characterised by critical exponents. These exponents are not independent but related by scaling relations, which we have obtained. The scaling approach is very powerful in characterising the critical behaviour in yielding the scaling laws, but does not lead to the actual values of the critical exponents. It is necessary to develop specific tools so that these critical exponents can be obtained. In this chapter we look at two approaches which allow for a quantitative calculation of the critical exponents. We start with the case of classical (thermal) transitions to introduce the methods.[1]

2.2 Landau Theory of Phase Transitions

The concept of order parameters plays a fundamental role in the theory of phase transitions. It characterises the phase with broken symmetry and the Landau free energy sufficiently close to the critical point can be expanded in powers of this parameter as

$$F_L = \int d^d x [a_0 + am^2(x) + um^4(x) - hm(x)], \qquad (2.1)$$

where $a = a_1 + a_2\epsilon + O(\epsilon^2)$, are analytic functions of the reduced temperature $\epsilon = (T - T_c)/T_c$. Above, a_0 represents the analytical background and will be set equal to zero. We will also use the notation $r_0 = a_2\epsilon$. In the most common case in condensed matter, $m(x)$, represents the magnetisation of a ferromagnet

---

[1] In Chapter 1, we distinguished the thermal exponents by a *tilde*. In this chapter we are mostly dealing with thermal phase transitions, and remove the *tilde* for simplicity. We will make clear in the text the nature of the exponents in each section.

at position x and the constant symmetry-breaking field h is a uniform magnetic field. Eq. 2.1 is an analytic expansion in terms of the order parameter, which only contains terms satisfying the symmetry of the ordered phase: in the present case, that of spin reversal such that $m \to -m$ and $h \to -h$. For a uniform magnet, the magnetisation can be taken independent of the position. In this case,

$$F_L = L^d[am^2 + um^4 - hm].$$

The actual value of the magnetisation is obtained by minimising the free energy with respect to m. When the external magnetic field, $h = 0$, we find from $\partial F_L / \partial m = 0$

$$m = 0 \text{ or } m = \sqrt{\frac{-a(T)}{2u}}.$$

For $T > T_c$ ($\epsilon > 0$) with $a_1 \equiv 0$, the real solution is $m = 0$. From the equation, $m(T < T_c) = \sqrt{\frac{-a_2(T-T_c)}{2uT_c}}$ and the definition of the critical exponents in Chapter 1, we can identify the exponent β in Landau theory to be $\beta = 1/2$.

The free energy is given by

$$F_L = L^d a_0, \quad T > T_c$$
$$F_L = L^d \left[a_0 - \frac{a_2^2 |T - T_c|^2}{4u} \right], \quad T < T_c \tag{2.2}$$

since $m(T > T_c) = 0$. Note, in this equation, the coefficient of the quartic term, u appearing in the denominator of the free energy in the ordered phase. This quartic interaction is a dangerously irrelevant interaction in Landau theory. It does not directly determine the critical exponents, as will be shown when we study the Gaussian theory below, but cannot be made equal zero, as is clear from the equations above, for m and F_L.

The specific heat is obtained from $C_V = -T \partial^2 F_L / \partial T^2$. It vanishes in the disordered phase, i.e. $C_V(T > T_c) = 0$ and in the ordered phase, we get, $C_V(T < T_c)/L^d = a_2^2 T/2u$. Then, the specific heat has a discontinuity at T_c, $\Delta C_V / L^d = a_2^2 T_c / 2u$ associated with a thermal critical exponent $\alpha = 0$. From the magnetic equation of state,

$$2am + 4um^3 = h, \tag{2.3}$$

we find, at the critical isotherm, $T = T_c$ ($\epsilon = 0$), that $m(h) \propto h^{1/3}$, from which we identify the critical exponent $\delta = 3$. The susceptibility

$$\chi = \left(\frac{\partial m}{\partial h} \right)_T = \frac{1}{2[2a + 12um^2(h)]},$$

where $m(h)$ is a solution of Eq. 2.3. For $h = 0$, in the cases $\epsilon < 0$ and $\epsilon > 0$, $\chi = 1/(8a_2\epsilon)$ and $\chi = 1/(4a_2\epsilon)$, respectively. In both cases the susceptibility diverges as $\chi \propto |T - T_c|^{-1}$, such that the critical exponent of the order parameter susceptibility is $\gamma = 1$.

We have obtained above the critical exponents of the Landau theory, $\alpha = 0$, $\beta = 1/2$, $\delta = 3$ and $\gamma = 1$, without taking into account details of the system, as the form of the interactions, the number of components of the order parameter or even the dimension of the system. In spite of this universal behaviour, experiments do not confirm the values of the critical exponents given by this theory. The reason, as we shall see, is the neglect of fluctuations. These will be partly taken into account below when we consider the Gaussian theory. However, we will have to wait until the introduction of the renormalisation group for a quantitative calculation of the critical exponents which yields agreement with experiments.

An important feature that is lacking in the Landau free energy above is to take into account spatial variations of the order parameter and a term that considers the energy cost of this variation. This is required to introduce a length scale in the problem so that we can define a correlation length and its associated critical exponent ν. In order to see how such term arises, we write the familiar Heisenberg Hamiltonian as

$$H = -\frac{1}{2}\sum_{i,j} J_{ij}\mathbf{S}_i.\mathbf{S}_j = -\frac{1}{2}\sum_i \int dr_{ij} J(r_{ij})\mathbf{S}(r_i)\exp(-r_{ij}.\nabla)\mathbf{S}(r_i),$$

with $\mathbf{S}(r_i) =< \mathbf{S} > +\mathbf{m}(r_i)$. If $J(r_{ij}) = J$, for $|r_i - r_j| < R$ and $J(r_{ij}) = 0$, otherwise, then

$$H \approx -\frac{pJR^2}{4a^d}\int d^d r |\nabla\mathbf{m}(r)|^2,$$

where p is the coordination number of the lattice and a the lattice spacing.

From now on, we consider as our basic functional for thermal phase transitions:

$$F_L = \int d^d x [|\nabla\mathbf{m}|^2 + r_0 m^2 + u m^4 - hm] \tag{2.4}$$

where, $r_0 = a_2\epsilon$ and

$$m^2 = m(x).m(x) = \sum_{i=1}^n [m_i(x)]^2$$

$$m^4 = [m^2]^2$$

$$|\nabla m|^2 = \sum_{\alpha=1}^d \sum_{i=1}^n \left(\frac{\partial m_i}{\partial x_\alpha}\right)^2$$

with n the number of spin components.

2.3 Gaussian Approximation ($T > T_c$)

As we pointed out before, the mean field approximation neglects fluctuations, as is clear from the fact that the specific heat is zero in the disordered phase. This leads to results for the critical exponents which are in disagreement with experiments in real magnets. The next step to improve this situation is to introduce a Gaussian approximation where fluctuations, but not their interactions, are taken into account. This corresponds to considering the following functional (in the absence of an external magnetic field):

$$F_L = \int d^d x [(\nabla m)^2 + r_0 m^2]$$

where $r_0 = a_2(T - T_c)/T_c$ with a_2 a constant. Since $m(r)$ can be written in terms of its Fourier components m_k, as $m(r) = \sum_k m_k exp(-ik.r)$, we find

$$F_L = \frac{1}{L^d} \sum_k \frac{1}{2} |m_k|^2 [k^2 + r_0].$$

The free energy F is obtained from

$$e^{-\beta F} = \int_{-\infty}^{+\infty} \prod_k d(\Re e m_k) d(\Im m m_k) e^{-\frac{\beta}{2}[k^2 + r_0]|m_k|^2}.$$

This is easily integrated to yield

$$F = -\frac{1}{2} k_B T \sum_{|k| < \Lambda} \ln \frac{2\pi L^d k_B T}{k^2 + r_0}.$$

This can be calculated by transforming the sum into an integral, using $\sum_k \rightarrow (L/2\pi)^d \int d^d k$ and Λ as an ultraviolet cut-off.

The fluctuations of the order parameter, $< |m_k|^2 >$, are obtained from

$$< |m_k|^2 > = \int_{-\infty}^{+\infty} \prod_k d(\Re e m_k) d(\Im m m_k) e^{-\frac{\beta}{2}[k^2 + r_0]|m_k|^2} |m_k|^2 = \frac{L^d k_B T}{k^2 + r_0}.$$

Since

$$< |m_k|^2 > = \int d^d x_1 d^d x_2 e^{ik.(x_1 - x_2)} < m(x_1)m(x_2) > \equiv L^d G(k)$$

with $< m(x_1)m(x_2) > = < m(x_1 - x_2)m(0) >$, the *correlation function* $G(k)$ is given by

$$G(k) \approx \frac{k_B T}{k^2 + \xi^{-2}},$$

where the correlation length, $\xi = (1/r_0)^{-1/2}$ diverges with the Gaussian or mean field exponent, $\nu = 1/2$. From the sum rule $G(k = 0) = k_B T \chi$, the static susceptibility is found to be given by, $\chi = 1/r_0$, yielding the Gaussian value of the susceptibility exponent $\gamma = 1$. Also from the k-dependence of the correlation function at the critical point, $r_0 = 0$, and the definition of the critical exponent η by, $G(k, r_0 = 0) \propto 1/(k^{2-\eta})$, we obtain the exponent $\eta = 0$, within the Gaussian approximation.

The heat capacity per unity volume, $C_V = -T(\partial^2 f/\partial T^2)$, with $f = F/L^d$, is given by

$$C_V/T = \frac{\partial^2}{\partial T^2} \left(\frac{k_B T}{2} \int\limits_0^\Lambda \frac{d^d k}{(2\pi)^d} Ln \frac{2\pi L^d k_B T}{k^2 + r_0} \right).$$

The dominant, most divergent contribution, which determines the critical behaviour of the specific heat, is given by

$$C_V \propto \int_0^\Lambda \frac{dk}{(2\pi)^d} \frac{k^{d-1}}{(r_0 + k^2)^2}.$$

In the process of extracting the divergencies, we put r_0 in evidence, define $x = k\xi$ with $\xi = 1/\sqrt{r_0}$, to get

$$C_V \propto \xi^{4-d} \int_0^{x_{max}} dx \frac{x^{d-1}}{(1 + x^2)^2},$$

where $x_{max} = \Lambda\xi$. For $d > 4$ the integral, which diverges for $\Lambda\xi \to \infty$, is compensated by the vanishing pre-factor to yield a finite result. For $d = 4$, we get a logarithmic divergence at T_c which is associated with an exponent $\alpha = 0$. For $d < 4$ the specific heat diverges at the transition. We write the final result as

$$C_V \propto \left| \frac{T - T_c}{T_c} \right|^{-(4-d)/2} \tag{2.5}$$

such that the Gaussian exponent $\alpha = (4 - d)/2$ depends on dimensionality differently from the mean field case where $\alpha = 0$ for any d. Note that the divergence of C_V arises from the lower limit of the integral, i.e. from long wavelength modes, since Λ is kept finite. Also, fluctuations give rise to a finite specific heat in the disordered phase, above the critical temperature, differently from naive mean field which yields $C_V(T > T_c) = 0$. Since the Gaussian correlation length exponent, $\nu = 1/2$, it turns out that Gaussian exponents obey the hyperscaling relation for thermal phase transitions for any dimension $d \leq 4$. In fact, if we substitute $\nu = 1/2$ in the classical hyperscaling relation, Eq. 1.3, we obtain the expression above for the exponent α of the Gaussian model.

Notice that for $d > 4$, fluctuations do not lead to a diverging specific heat at the transition. On the other hand, for $d = 4$, $\alpha_{gaussian} = \alpha_{mean-field} = 0$. These simple results are the essence of the concept of *upper critical dimension* and substantiate statements, as *above the upper critical dimension the critical exponents are mean field-like*. Indeed, they imply that for $d \geq 4$, fluctuations are irrelevant and consequently Landau theory, which neglects these fluctuations, yields the correct description of the critical behaviour. Then $d_c = 4$ is the *upper critical dimension* for the magnetic transitions we have been describing. Note that, for quantum phase transitions, since the effective dimension is $d_{eff} = d + z$, it turns out that the upper critical dimension is reached at lower Euclidean dimensions d than for thermal transitions.

There is an interesting point related to the classical hyperscaling relation, $2 - \alpha = \nu d$, which is worth pointing out. Let us substitute the mean field exponents, $\nu = 1/2$ and $\alpha = 0$, in this equation. We find that it is satisfied for $d = d_c = 4$. This is a general feature of phase transitions, that the hyperscaling relation is satisfied by the mean field exponents just at the upper critical dimension. Toulouse (1974) used this fact to obtain the upper critical dimension for the percolation problem. He argued that the solution for the percolation problem in the Bethe lattice represents the mean field solution of this problem. The relevant critical exponents can be exactly obtained in this case and, when substituted in the hyperscaling relation, it is satisfied for $d = d_c^{perc} = 6$. Most of the problems we consider in this book have an upper critical dimension. This is certainly the case for the Kondo lattice. However, this is not a general rule. In disordered systems, for example, the upper critical dimension may be infinite, such that the critical exponents depend on dimensionality for any d.

Breakdown of Hyperscaling Above d_c

Consider the expression for specific heat in the Gaussian approximation given by Eq. 2.5. The specific heat exponent $\alpha = (4 - d)/2$ becomes zero for $d = d_c$ and negative for $d > d_c = 4$. This implies that, for $d \geq d_c$, the singularity of the specific heat in the Gaussian approximation becomes weaker that in the mean field approach. However, this is unphysical since this singularity cannot become weaker than in mean field, which does not consider fluctuations. To make sense of this simple observation, we have to assume that the critical exponent α remains fixed at the mean field value $\alpha = 0$ for $d \geq d_c$. From the perspective of hyperscaling this implies that the relation $2 - \alpha = \nu d$ has to be violated for $d > d_c$ since $\alpha = 0$ and fixed for all $d > d_c$. We will have the opportunity to discuss this in much more detail in the next chapters.

Ginzburg Criterion

As shown above, for dimensions $d \geq 4$,[2] Landau theory gives a correct description of critical behaviour since fluctuations become irrelevant. What happens in lower dimensions? Is there a region sufficiently away from the critical temperature where the mean-field theory is valid? For a quantitative estimation of the importance of fluctuations, we calculate the ratio $\Delta \bar{m}^2 / \bar{m}^2$, where the average $\bar{x}$ is taken over a region of the size of the correlation length. This quantity can be evaluated as (Goldenfeld, 1992)

$$\frac{|\int G(r) dr|}{|\int dr m^2(r)|}, \tag{2.6}$$

and it can be much larger than unity when fluctuations become important. The function $G(r)$ is the order parameter correlation function defined previously. The sum rule for the susceptibility gives

$$\int G(r) dr \approx k_B T \chi.$$

On the other hand, the integral in the denominator of Eq. 2.6 is obtained as $\int dr m^2(r) \approx \xi^d m^2$. The correlation length we write as $\xi = R|\epsilon|^{-1/2}$, where the characteristic length R reflects the range of the interactions. Using the expressions for the susceptibility χ, the magnetisation m and the specific heat jump ΔC given by the Landau theory of Section 2.2, the condition for the ratio, Eq. 2.6, to be *much smaller* than unity, can be written as

$$\left| \frac{T - T_c}{T_c} \right|^{(4-d)/2} >> \frac{1}{(\Delta C / k_B)(R/a)^d}, \tag{2.7}$$

where a is an atomic distance. In the temperature range for which this inequality is satisfied, i.e. sufficiently far from T_c, the singularities are described by the Landau theory. When the phase transition is approached and fluctuations start to grow, the critical exponents deviate from their Landau or mean field values. Notice that the presence of the characteristic length R in the denominator of the expression above implies that the longer the range of the interactions, the closer to the transition the mean field exponents remain valid. This is the case, for example, of BCS superconductors which have a mean field-like transition due to the large sizes of Cooper pairs. Using $\Delta C / k_B \approx 1$ per particle and $(R/a) \approx 10^3$ for a Cooper pair one can see that the number on the right-hand side of Eq. 2.7 can become very small. In Section 2.5, we study a model with infinite range interactions for which mean field theory yields the exact solution.

---

[2] The upper critical dimension $d = 4$ is known as the *marginal case* and is generally accompanied by logarithmic singularities, as obtained for the specific heat using the Gaussian free energy for $d = 4$.

2.4 Gaussian Approximation ($T < T_c$)

We shall now extend the Gaussian approximation to below the critical temperature where there is a finite-order parameter in the system. We consider again the magnetic case. Below T_c, we have to distinguish fluctuations along and perpendicular to the direction of the order parameter. We start with the Landau–Ginzburg (LG) functional:

$$F_L = \int d^d x [(\nabla m)^2 + r_0 m^2 + u m^4 - hm].$$

Minimisation of this functional leads to the following equation for $\overline{m}$, the space-independent average magnetisation:

$$\overline{m}(2r_0 + 4u\overline{m}^2) - h = 0.$$

The local magnetisation, site-dependent magnetisation can be written as

$$m(x) = \overline{m}[\mathbf{n} + \delta m_1(x)\mathbf{n} + \boldsymbol{\delta m}_\perp(x)]$$
$$= \overline{m}(1 + \delta m_1)\mathbf{n} + \overline{m}\boldsymbol{\delta m}_\perp$$

where $\mathbf{n}$ is a unit vector in the direction 1 of the spontaneous magnetisation, which coincides with that of the external magnetic field $\mathbf{h}$. The fluctuations in the directions perpendicular to that of the order parameter are given by $\boldsymbol{\delta m}_\perp$. We can write the terms in LG functional in this representation. We get

$$m^2(x) = \overline{m}^2(1 + \delta m_1)^2 + \overline{m}^2|\boldsymbol{\delta m}_\perp|^2$$
$$= \overline{m}^2(1 + 2\delta m_1 + \delta m_1^2 + \delta m_\perp^2),$$

where $\delta m_\perp^2 = |\boldsymbol{\delta m}_\perp|^2$. The quartic term to order $O(\delta m)^2)$ is given by

$$m^4(x) \approx \overline{m}^4(1 + 4\delta m_1 + 6\delta m_1^2 + 2\delta m_\perp^2).$$

The Landau–Ginzburg free energy density, to $O(\delta m^2)$, becomes

$$F_L(x) = \overline{m}^2[(\nabla \delta m_1)^2 + (\nabla \delta m_\perp)^2 + r_0\overline{m}^2(1 + 2\delta m_1 + \delta m_1^2 + \delta m_\perp^2)$$
$$+ u\overline{m}^4(1 + 4\delta m_1 + 6\delta m_1^2 + 2\delta m_\perp^2) - h\overline{m}(1 + \delta m_1),$$

or

$$F_L(x) = \overline{m}^2[(\nabla \delta m_1)^2 + (\nabla \delta m_\perp)^2] + (r_0\overline{m}^2 + u\overline{m}^4 - h\overline{m})$$
$$+ \delta m_1[2r_0\overline{m}^2 + 4u\overline{m}^4 - h\overline{m}] + \delta m_1^2[r_0\overline{m}^2 + 6u\overline{m}^4]$$
$$+ \delta m_\perp^2[r_0\overline{m}^2 + 2u\overline{m}^4].$$

Using the minimisation condition $\overline{m}(2r_0 + 4u\overline{m}^2) - h = 0$, or $2r_0\overline{m}^2 + 4u\overline{m}^4 - h\overline{m} = 0$, we obtain

$$F_L(x) = \overline{m}^2[(\nabla \delta m_1)^2 + (\nabla \delta m_\perp)^2] + (r_0\overline{m}^2 + u\overline{m}^4 - h\overline{m})$$
$$+ \delta m_1^2[4u\overline{m}^4 + h\overline{m}/2] + \delta m_\perp^2[h\overline{m}/2],$$

or

$$F_L(x) = F_{L0} + \overline{m}^2\{[(\nabla\delta m_1(x))^2 + (\nabla\delta m_\perp(x))^2]$$
$$+ \delta m_1^2(x)[4u\overline{m}^2 + h/2\overline{m}] + \delta m_\perp^2(x)[h/2\overline{m}]\},$$

where F_{L0} is the mean field, space-independent contribution. The Landau–Ginzburg functional, $F_L = \int F_L(x)d^dx = L^d F_{L0} + F_G$. Fourier transforming the second, Gaussian part, we get

$$F_G = \overline{m}^2 \int d^dk\{[k^2 + 4u\overline{m}^2 + h/2\overline{m}]|\delta m_1(k)|^2 + [k^2 + h/2\overline{m}]|\delta m_\perp|^2\}.$$

The contribution of transverse excitations to this functional in the absence of an external magnetic field vanishes for $k \to 0$. This is associated with the divergence of the transverse correlation function $G_\perp(k) = <|m_\perp(k)|^2>$ in this limit and is a consequence of an important result, namely, the Goldstone theorem. Since the ferromagnetic ground state breaks the rotational invariance of the Landau–Ginzburg functional, Goldstone's theorem predicts the existence of a zero–energy mode which restores the broken symmetry. The divergence of the transverse correlation function at $k = 0$ is due to the infinite range correlation of the spins performing a uniform rotation. In the next section we further investigate Goldstone's result using a microscopic model.

Another feature worth pointing out in the equation above for F_G is that the quantity $\overline{m}^2$ appears multiplying the k^2 term in the transverse part of the functional. This quantity measures the rigidity of the transverse modes and plays the role of the *stiffness* ρ_{sw} of the transverse, *spin waves* excitations. In the disordered phase longitudinal and transverse excitations are indistinguishable. However, in the broken symmetry phase, as the critical point is approached from the ordered phase, the longitudinal correlations behave similarly to the correlations in the disordered phase, but the transverse ones have a quite different dynamic behaviour. The stiffness ρ_{sw} of the spin waves decreases as the transition is approached, as $\rho_{sw} \propto |\epsilon|^{2\beta}$ and vanishes at the critical point. In a more general approach the critical behaviour of the stiffness, which is the coefficient of the $|\nabla m|^2$ in the Landau functional is obtained as follows. Let us consider this part of the functional, i.e.

$$L_\nabla = \rho_{sw} \int d^dx |\nabla m|^2. \tag{2.8}$$

From dimensional analysis, we expect that under a length scale transformation by a factor b, ρ_{sw} should scale as

$$\rho'_{sw} = b^{d-2}\rho_{sw}. \tag{2.9}$$

Using the results of Chapter 1, we find that close to the transition, the stiffness behaves as $\rho_{sw} \propto |\epsilon|^{\nu(d-2)}$. Notice that for $d = 1$, ρ_{sw} scales to zero under repeated scale transformations. This implies that there is no long range-ordered magnetic phase in a one-dimensional classical system with continuous symmetry, even at $T = 0$. The case $d = 2$ is marginal and is such that at $T = 0$ there may be long range magnetic order but not at finite T. The concept of magnetic stiffness is particularly useful for systems at the lower critical dimension as the $2d$ XY model. In this case even though there is no long-range magnetic order at $T \neq 0$, the stiffness may be finite and this distinguishes this system from a common paramagnetic system. In the next chapters we will explore the concept of stiffness in different contexts.

In the quantum case, anticipating the results of the next chapters, the relevant quantity describing the system near the quantum critical point is the *effective action*,

$$S_{eff} = \int d^d x \int dt \mathcal{L}(x, t),$$

where $\mathcal{L}$ is the Lagrangian density. The derivative part of the Lagrangian, with respect to space-time coordinates, also allows a generalised stiffness for transverse excitations to be defined, which, in the quantum case, scales as

$$\rho'_{sw} = b^{d+z-2} \rho_{sw}$$

such that, close to the critical point, this behaves as $\rho_{sw} \propto |g|^{\nu(d+z-2)}$.

2.5 Goldstone Mode

Let us consider a spin-1/2 Heisenberg ferromagnet with infinite range interactions, described by the Hamiltonian

$$H = -\frac{1}{2} \sum_{i \neq j} J_{ij} \mathbf{S_i}.\mathbf{S_j} \tag{2.10}$$

with $J_{ij} = J/N$. This model is exactly soluble by the mean field theory (Stanley, 1971). It has a continuous symmetry $O(n)$ where n is the number of components of the spins which we take equal to the space dimension. The ferromagnetic state breaks rotational symmetry and we expect the existence of a Goldstone mode, i.e. a boson with zero energy which restores the broken rotational invariance of the Hamiltonian. Let us look at the dynamics of this model calculating the transverse dynamic susceptibility or *Green's function* defined by (see Section A.1 on Green's functions at the end of this chapter):

$$G_{ij}^{+-}(t) = << S_i^+(t) | S_j^-(0) >> = i\theta(t) < [S_i^+(t), S_j^-(0)] >,$$

where S_i^+, S_j^- are spin-raising and -lowering operators, respectively. The operators are written in the Heisenberg representation, $\theta(x)$ is the *theta* function and the average $< \cdots >$ is taken over a grand canonical ensemble. The equation of motion for this Green's function is given by

$$i\frac{dG^{+-}_{ij}(t)}{dt} = -\delta(t) < [S^+_i(t), S^-_j(0)] > + << [S^+_i(t), H]|S^-_j(0) >> .$$

Calculating the commutator of S^+ with the Hamiltonian, Eq. 2.10, we obtain:

$$i\frac{dG^{+-}_{ij}(t)}{dt} = -2\delta(t) < S^z > \delta_{ij} + \sum_l J_{il} << S^z_l S^+_i(t)|S^-_j(0) >>$$

$$- \sum_l J_{il} << S^z_i S^+_l(t)|S^-_j(0) >> .$$

In the case of infinite range interactions, the following decoupling becomes exact:

$$<< S^z_l S^+_i(t)|S^-_j(0) >> = < S^z > << S^+_i(t)|S^-_j(0) >> .$$

Using this decoupling, the equation of motion for the Green's function can be formally integrated to yield

$$G^{+-}_{ij}(t) = 2i\sigma\theta(t) \left(\exp(i\sigma t(\hat{\Omega}_1 + \hat{\Omega}_2)) \right)_{ij}, \qquad (2.11)$$

where we wrote $\sigma = < S^z >$. The matrices $\hat{\Omega}_1$ and $\hat{\Omega}_2$ are given by

$$\hat{\Omega}_1 = J/N$$
$$\hat{\Omega}_2 = -J\delta_{ij}.$$

Since the diagonal matrix $\hat{\Omega}_2$ commutes with $\hat{\Omega}_1$, we can rewrite Eq. 2.11 as

$$G^{+-}_{ij}(t) = 2i\sigma\theta(t)\exp(-i\sigma t J) \left(\exp(i\sigma t\hat{\Omega}_1) \right)_{ij}.$$

The matrix element

$$\left(\exp(i\sigma t\hat{\Omega}_1) \right)_{ij} = \sum_\lambda a_{\lambda i}a_{\lambda j}\exp(i\sigma t\theta_\lambda),$$

where the θ_λs are the eigenvalues of $\hat{\Omega}_1$ and $a_{\lambda i}$ is the ith component of the λth eigenvalue. Since the eigenvalues are normalised, $|a_{\lambda i}|^2 = 1/N$, the local transverse susceptibility can be written as

$$G^{+-}_{ii}(t) = 2i\sigma\theta(t)\exp(-i\sigma t J)\frac{1}{N}\sum_\lambda \exp(i\sigma t\theta_\lambda).$$

The density of eigenvalues of the $N \times N$ matrix $\hat{\Omega}_1$ with all elements equal to J/N is given by

$$\rho(\theta_\lambda) = \frac{N-1}{N}\delta(\theta_\lambda) + \frac{1}{N}\delta(\theta_\lambda - J).$$

In the limit of very large N, we can substitute the sum over eigenvalues by an integral over the density $\rho(\theta_\lambda)$, i.e. $(1/N)\sum_\lambda \to \int d\theta_\lambda \rho(\theta_\lambda)$ to get

$$G_{ii}^{+-}(t) = 2i\sigma\theta(t)\left[\frac{N-1}{N}\exp(-i\sigma t J) + \frac{1}{N}\right].$$

Fourier transforming in time yields

$$G_{ii}^{+-}(\omega) = -\frac{1}{\pi}\left[\frac{N-1}{N}\frac{\sigma}{w - \sigma J + i\epsilon} + \frac{1}{N}\frac{\sigma}{\omega + i\epsilon}\right],$$

from which we get the spectral density:

$$\Im m G_{ii}^{+-}(\omega) = \frac{N-1}{N}\delta(\omega/\sigma - J) + \frac{1}{N}\delta(\omega/\sigma). \tag{2.12}$$

Thus the spectrum of transverse excitations consists of $N-1$ modes of energy σJ and a mode of zero energy that can be identified with the symmetry restoring, or Goldstone, mode (Continentino and Rivier, 1977). This zero energy excitation is associated with a rotation of the entire system and restores the original $O(3)$ symmetry of the Hamiltonian broken by the appearance of ferromagnetic order.

Using the identity $\sigma = \langle S^z \rangle = 1/2 - \langle S^- S^+ \rangle$, for spin-1/2 and the fluctuation-dissipation theorem to calculate the correlation function $<S^- S^+>$ from the transverse susceptibility, we get the mean field equation for the magnetisation of the ferromagnet, $\sigma = \tanh(\beta\sigma J)$, as expected. There is also a singular contribution arising from the Goldstone mode. This is dealt with, applying an infinitesimal external magnetic field in the system and taking the thermodynamic limit.

For nearest neighbour interactions, the excitations of the Heisenberg ferromagnet are spin-waves, which in the hydrodynamic limit have a dispersion relation $\omega = Dk^2$. The Goldstone mode in this case is the $k = 0$, zero energy, spin-wave mode which corresponds to a uniform rotation of the whole system.

2.6 Ising Model in a Transverse Field – Mean-Field Approximation

As an illustration of the mean field approximation to a quantum problem, let us consider the Ising model in a transverse field:

$$H - \sum_{ij} J_{ij} S_i^z S_j^z - h \sum_I S_i^x.$$

We write the operators for the spin-1/2 in terms of Pauli matrices, $\vec{S} = (1/2)\vec{\sigma}$. The mean field Hamiltonian for nearest neighbour interaction J with p neighbours is given by

$$H_{MF} = -\frac{pJ <S^z>}{2}\sum_i \sigma_i^z - \frac{h}{2}\sum_i \sigma_i^x,$$

which is a two-by-two matrix. The eigenvalues are given by

$$\lambda = \pm\frac{1}{2}\sqrt{(pJ < S^z >)^2 + h^2}.$$

The self-consistent equation for the average magnetisation is

$$< S^z > = \frac{Tr S^z e^{-\beta H}}{Tr e^{-\beta H}} = \frac{pJN < S^z >}{2\sqrt{p^2 J^2 < S^z >^2 + h^2}} \tanh[\frac{\beta}{2}\sqrt{p^2 J^2 < S^z >^2 + h^2}],$$

where the trace was calculated on the basis where H_{MF} is diagonal. At $T = 0$, the self-consistent equation for the magnetisation is

$$< S^z > = \frac{N}{2}\frac{pJ < S^z >}{\sqrt{p^2 J^2 < S^z >^2 + h^2}}.$$

This equation yields, at $T = 0$, a magnetisation $m = < S^z >$ that vanishes at the critical value of the ratio, $(h/J)_c = p$ as $m \propto |g|^{1/2}$ where $|g| = |(h/J) - (h/J)_c|$. Then the order parameter exponent at the zero temperature transition takes the mean field value $\beta = 1/2$. There is also a line of finite temperature instabilities for $(h/J) < (h/J)_c$ where $< S^z(T) >$ vanishes. This line, close to the zero temperature critical point, is given by $T_c \propto |g|^{1/2}$, such that the shift exponent takes the mean field value $\psi = 1/2$. One can easily check that in the approximation used above all critical exponents are mean field-like for both the zero and finite temperature transitions. All these results for the critical exponents can be obtained using a Landau–Ginzburg functional given by:

$$F_0 = (|g| - T^2)m^2 + um^4.$$

What about the dynamic exponent z? In the simple mean-field approximation presented here there is no way of identifying the exponent z. There are no dynamics in this mean field approach. The non-critical side of the phase diagram in this approximation is also trivial and uninteresting with no crossover phenomena. This is not necessarily the case at this level of approximation. There are more sophisticated mean field theories, which take into account correlations between pairs and yield non-trivial results in the disordered phase as new energy scales and crossover temperatures, even in quantum problems (Granato and Continentino, 1993). There are also dynamical mean field theories in which it makes sense to define a dynamic exponent, as will be shown in Chapter 10.

A more sophisticated approach to the Ising model in a transverse field has been given by A.P. Young (Young, 1975) in a pioneering work. An essential point of his method is to take into account the non-commutativity of the two competing terms in the Hamiltonian. He uses the basic identity

$$Tr e^{-\beta(H_0 + H_1)} = Tr e^{-\beta H_0} T e^{-\int_0^\beta H_1(\tau)d\tau}, \tag{2.13}$$

where $H_1(\tau) = e^{\tau H_0} H_1 e^{-\tau H_0}$ and T is the *time ordering* operator, to introduce the *dynamics* in the problem. Young was able to identify the dynamic exponent $z = 1$ for the transverse field Ising model (TFIM), such that the quantum transition in $d = 1$ has the same critical exponents obtained by Onsager for the finite temperature transition of the two dimensional classical Ising model (Pfeuty, 1970). An interesting feature of the mean field solution given here for the TFIM is that it provides the exact critical exponents for the quantum phase transition of this model, excluding z, of course, in $d = 3$. The reason, as will be discussed many times in this book, is that the effective dimension for this problem $d_{eff} = d + z$. In $d = 3$ with $z = 1$ we get $d_{eff} = 4$, which is the same as the upper critical dimension $d_c = 4$ for this type of phase transition.

3

Real Space Renormalisation Group Approach

3.1 Introduction

The real space renormalisation group has a wide application in the study of classical critical phenomena. For quantum systems, several versions of the real space renormalisation group have been proposed (Jullien, 1981; Pfeuty *et al.*, 1982; dos Santos, 1982). The real space *RG* is the most direct way of implementing the scaling ideas presented in Chapter 1. The operation of elimination of degrees of freedom as the system is reduced to blocks of smaller sizes under a change of length scale is directly and intuitively represented as mathematical equations. In particular, the *block method* has been used to study different problems of strongly correlated systems (Pfeuty *et al.*, 1982) and spin chains (White, 1992, 1993) with different degrees of sophistication. We show here in detail a simple form of this technique applied to the case of the one-dimensional Ising model in a transverse field. The purpose is to illustrate the ideas introduced earlier in Chapter 1 with a concrete example where a quantum critical point is associated with an unstable fixed point. Also, we show how the stable attractors or fixed points can characterise the different phases and how the critical exponents can be obtained. Although the actual values for the quantum critical point and critical exponents that we obtain using a small block are not exact, a procedure is outlined which can be used to improve these results. In this sense, the block method presented here represents a controlled approximation, which can be systematically improved to yield in the limit of very large blocks essentially the correct results. When compared to the *k*-space version of the RG (to be presented in the next two chapters), the real space version turns out to be in general more difficult to control, in the sense of producing a systematic improvement of the results. On the other hand, this method yields without much effort the attractive fixed points of the different phases, allowing for their characterisation. This turns out to be useful since the nature of the phase transition, whether first or second order, can be determined by exponents calculated in the attractor of this phase (Nienhuis and Naunberg, 1975).

3.2 The Ising Model in a Transverse Field

The Hamiltonian describing an infinite chain of $S = 1/2$ Ising spins in a transverse field was given in Chapter 1. Here we rotate it for computational simplicity and write

$$H = -J \sum_{i,j} S_i^x S_j^x - h \sum_i S_i^z. \tag{3.1}$$

In the absence of the transverse field h, the Ising chain is ordered at $T = 0$ with $< S^x > \neq 0$. In the RG block method the idea is to divide the chains in blocks of the same arbitrary size, which are going to be treated exactly. We then take the two lower energy states of this block, which are going to be considered as the two states of a new effective spin $1/2$. The coupling between these effective spins is than calculated. We study here the simple case of a block of two spins which corresponds to a scale transformation by a factor $b = 2$. Let us write the Hamiltonian for such a block:

$$H = -J S_1^x S_2^x - h(S_1^z + S_2^z). \tag{3.2}$$

In order to find the eigenvalues of this block, we choose the basis of eigenstates of the operators S^z, i.e. $|+ >, |- >$. We introduce *raising* and *lowering* operators, defined by

$$S^+ = (1/2)(S^x + i S^y)$$
$$S^- = (1/2)(S^x - i S^y)$$

such that $S^x = S^+ + S^-$. On the other hand,

$$S^+|+ >= 0 \ S^+|- >= |+ > \ S^z|+ >= +1|+ > \tag{3.3}$$
$$S^-|- >= 0 \ S^-|+ >= |- > \ S^z|- >= -1|- > . \tag{3.4}$$

The Hamiltonian matrix of the block, written in the basis formed by the product of the eigenstates of $S_1^z S_2^z$, $(|+ + >= |+ >_2 |+ >_1, |- - >= |- >_2 |- >_1, |+ - >= |+ >_2 |- >_1$, and $|- + >= |- >_2 |+ >_1)$, is given by

$$H_{12} = \begin{pmatrix} -2h & -J & 0 & 0 \\ -J & 2h & 0 & 0 \\ 0 & 0 & 0 & -J \\ 0 & 0 & -J & 0 \end{pmatrix}. \tag{3.5}$$

This matrix is block-diagonal and can easily be diagonalised to yield the eigenvalues. These are found to be $\varepsilon_0 = -\sqrt{4h^2 + J^2}$, $\varepsilon_1 = -J$, $\varepsilon_2 = +J$ and $\varepsilon_3 = +\sqrt{4h^2 + J^2}$. The corresponding eigenvectors $|i >$, such that $H_{12}|i >= \varepsilon_i |i >$, are given by

$$|0>= \frac{1}{\sqrt{1+d_1^2}} [d_1|++>+|-->]$$ (3.6)

$$|1>= \frac{1}{\sqrt{2}} [|+->+|-+>]$$ (3.7)

$$|2>= \frac{1}{\sqrt{2}} [|+->-|-+>]$$ (3.8)

$$|3>= \frac{1}{\sqrt{1+d_2^2}} [(-d_2)|++>+|-->],$$ (3.9)

where

$$d_1 = \frac{J}{\sqrt{4h^2+J^2}-2h}$$ (3.10)

and

$$d_2 = \frac{J}{\sqrt{4h^2+J^2}+2h}.$$ (3.11)

Next step, we take the two lower energy states, $(|0>, |1>)$ as the two states of a fictitious spin $S' = 1/2$ which represents the effective spin of the block, i.e. we take

$$|0>\rightarrow |+'>$$ (3.12)

$$|1>\rightarrow |-'>.$$ (3.13)

Then the new Hamiltonian of the block is given by

$$H_0' = \varepsilon_0|+'><+'|+\varepsilon_1|-'><-'| \equiv -h'S^{z'}+C',$$ (3.14)

and the last equation defines the renormalised transverse field h' and the constant C'. Now,

$$H_0'|+'>= \varepsilon_0|+'>= (-h'+C')|+'>$$ (3.15)

$$H_0'|-'>= \varepsilon_1|-'>= (+h'+C')|-'>$$ (3.16)

such that

$$h' = \frac{\varepsilon_1-\varepsilon_0}{2} = \frac{\sqrt{4h^2+J^2}-J}{2}$$ (3.17)

and

$$C' = \frac{\varepsilon_1+\varepsilon_0}{2} = \frac{\sqrt{4h^2+J^2}+J}{2}.$$ (3.18)

The first of these equations is the recursion relation for the renormalised transverse magnetic field acting on the new effective spins replacing the blocks. We now calculate the coupling between these effective spins. The easiest way to do that is to write the x-component of an original spin, S_i^x in terms of that of the block effective spin $S_b'^x$, i.e. $S_i^x = \xi S_b'^x$. Above, $i = 1$ or $i = 2$, either one of the two spins joining to make the block $b = 12$. The quantity ξ is obtained by calculating the matrix element:

$$< +'|S_i^x|-' > = \xi < +'|S_b'^x|-' > \equiv \xi, \qquad (3.19)$$

where $|+' > = |0 >$ and $|-' > = |1 >$, with $|0 >$ and $|1 >$ the eigenvectors, corresponding to the two smallest eigenvalues ε_0 and ε_1, respectively. We get

$$\xi = \frac{1 + d_1}{\sqrt{2(1 + d_1^2)}}$$

independent of a particular block. Defining the coupling J' between blocks 12 and 34, made of spins 1 and 2 and 3 and 4 respectively, through the equation

$$J' S_{12}'^x S_{34}'^x = J S_2^x S_3^x$$

we obtain, using the results above, $J' = J\xi^2$, or

$$J' = J \frac{(1 + d_1)^2}{2(1 + d_1^2)}, \qquad (3.20)$$

where d_1 was given before.

3.3 Recursion Relations and Fixed Points

The final recursion relations are given by

$$h' = \frac{\sqrt{4h^2 + J^2} - J}{2} \qquad (3.21)$$

$$J' = \frac{J}{4} \frac{[J - 2h + \sqrt{4h^2 + J^2}]^2}{[4h^2 + J^2 - 2h\sqrt{4h^2 + J^2}]}. \qquad (3.22)$$

Alternatively, we can introduce a variable $x = (h/J)$ in terms of which we get a single recursion relation given by

$$x_{n+1} = \frac{2[\sqrt{1 + 4x_n^2} - 1][1 + 4x_n^2 - 2x_n\sqrt{1 + 4x_n^2}]}{[1 + \sqrt{1 + 4x_n^2} - 2x_n]^2}. \qquad (3.23)$$

This equation is to be iterated starting with x_0, from which we obtain x_1, which in turn, when substituted in the equation above, yields x_2 and so on. The sequence of points $\{x_0, x_1, x_2, \cdot, x_n\}$ yields the *flow* of the RG equations. Physically, this sequence represents the continued operation of gathering the spins of the infinite chain in blocks and substituting them by new effective spins, as was done above.

Each time this operation is repeated, the length scale of the system changes by a factor $b = 2$, such that $L' = L/b$. The fixed points of the recursion relation $x_{n+1} = f(x_n)$ are those values of x for which $x_{n+1} = x_n$, i.e. they are the roots of the equation $x_c = f(x_c)$. This equation can be written as

$$8x_c^3 + 4x_c^2(1+4x_c^2)^{1/2} - 2(1+4x_c^2)^{3/2} + 4x_c^2 + 2x_c(1+4x_c^2)^{1/2} + 2x_c + 2 = 0. \quad (3.24)$$

It can easily be verified that $x_c = 0$ and $x_c = \infty$ are solutions of this equation. There is another non-trivial solution which has to be found numerically and is given by $x_c = 1.2767$. The nature of these *fixed points* is quite different. For the first two, if we start with a x_0 value not too far from them (actually from a value x_0 larger or smaller than $x_c = (h/J)_c = 1.2767$), the sequence of points $\{x_1, x_2, \cdots, x_n\}$ eventually converges to one of these fixed points. These are then *stable fixed points* or *attractors*. The one at $x_c = (h/J) = 0$ is clearly the strong coupling attractor of the ordered phase. The other at $(h/J) = \infty$ is the attractor of the disordered phase. On the other hand, starting from x_0 close to $x_c = 1.2767$, the flow of the recursion relations, i.e. the sequence $\{x_1, x_2, \cdots, x_n\}$, takes the system away from this *unstable* fixed point. It is with this unstable fixed point that the RG associates the quantum critical point separating the long range ordered magnetic phase, with $< S^x > \neq 0$, from the disordered phase.

Close to the unstable fixed point, $x_c = (h/J)_c = 1.2767$, we can expand the recursion relation, in the form we used earlier in Chapter 1:

$$x_{n+1} = x_c + b^a(x_n - x_c),$$

where $b = 2$ is the scaling factor associated with the particular block renormalisation we have used (a two-spins block). We recognise from our previous analysis the correlation length exponent $\nu = 1/a$, such that $\xi \propto |(h/J) - (h/J)_c|^{-\nu}$. Numerically, this exponent can easily be obtained from the expression

$$\nu = \frac{\ln b}{\ln(\frac{x_{n+1}-x_c}{x_n-x_c})} = \frac{\ln b}{\ln(\frac{\partial x'}{\partial x})_{x=x_c}}.$$

The last equation arises since x_n can be taken arbitrarily close to x_c. The present block RG with $b = 2$ yields $\nu = 1.47$ for the correlation length exponent, which should be compared with the exact result $\nu = 1$ (the same as the finite temperature $d = 2$ Ising model).

The dynamical exponent z can be obtained, noticing that *at the unstable fixed point* $(h/J)_c$, the recursion relation for the coupling can be written as

$$J' = b^{-z}J = J\frac{\sqrt{1+4x_c^2}-1}{2x_c}. \quad (3.25)$$

Since the dynamic exponent z controls the renormalisation of the coupling *at the fixed point*, as we saw previously in Chapter 1, we get for the dynamic exponent

$$z = \frac{\ln \frac{2x_c}{\sqrt{1+4x_c^2}-1}}{\ln b}. \tag{3.26}$$

Eq. 3.25 is specially interesting since it relates the dynamic exponent directly to renormalisation of the interaction at the QCP. As we will show in Chapter 5, the dynamic exponent in the frequency-momentum formulation of the RG arises from the time or frequency dependence of the action. In both cases, however, the underlying mathematical reason for the coupling between static and dynamics in quantum phase transitions is the non-commutativity of the competing terms in the Hamiltonian. From a more physical perspective we can trace it to the Heisenberg uncertainty relations.

Using the value of $x_c = 1.2767$ and $b = 2$, we get $z = 0.55$, to be compared with the exact result $z = 1$. Alternatively, z can be obtained from the scaling of the transverse field at the unstable fixed point since there also, $h' = b^{-z}h$. Then, in the real space renormalisation group approach, the dynamic exponent z is obtained exactly in the same way as it was introduced in the scaling theory.

The one-dimensional Ising model in a transverse field has been solved exactly by Pfeuty (1970), which allows a comparison of the block method with the exact results. The quantum critical point occurs at $(h/J)_c = 1$, instead of the result, $(h/J)_c = 1.2767$, obtained here. Furthermore, the exact values for the exponents are $z = 1$ and $\nu = 1$. The different values obtained by the block method can be attributed to the small, finite size of the blocks used here. In fact, the method above can be improved by increasing the size of the blocks. For blocks of five spins, such that $b = 5$, the quantum critical point is found at $(h/J)_c = 1.0797$ and $z = 0.705$ (Jullien, 1981): much closer to the exact results.

Since the exact value for the dynamic exponent is $z = 1$, this makes the effective dimension of the present problem, $d_{eff} = d + z = 1 + 1 = 2$. The exact exponents associated with the quantum critical point at $(h/J)_c$ in fact do coincide with those of the thermal phase transition of the classic two-dimensional Ising model found by Onsager.

3.4 Conclusions

In this chapter, we have presented a renormalisation group technique in real space which shows explicitly how some of the exponents introduced before can be found in a real space calculation. In particular, we have shown how the correlation length and the dynamic exponent can be obtained. Although the block method used here yields only approximate values for the critical point and exponents, it can be systematically improved to yield better results.

The block method can be generalised to finite temperatures, although for the problem above there is not much interest, since in $d = 1$ there is no line of finite temperature phase transitions. Real space renormalisation group techniques have been introduced to deal with hierarchical lattices at finite temperatures (dos Santos, 1982). These lattices provide a good approximation for Euclidean systems and can also be used to treat quantum disordered systems quite successfully (Boechat *et al.*, 1994).

4

Renormalisation Group: the ϵ-Expansion

4.1 The Landau–Wilson Functional

The Landau and Gaussian approaches presented in the previous chapters allow values for the critical exponents that govern the divergencies of physical quantities close to a phase transition to be obtained. These exponents obey scaling relations and they appear in the scaling form of the free energy close to a critical point. As we pointed out, the mean field Landau theory does not consider fluctuations of the order parameter and can be at most an approximation in a region of the phase diagram not too close to the transition, specified by the Ginzburg criterion. The Gaussian free energy considers fluctuations but neglects their interactions, which appear on the fourth order in the expansion of the free energy in terms of the order parameter. In this chapter we use the renormalisation group (RG) (Anderson, 1970; Wilson, 1971, 1975; Nogueira, 1996) to treat the complete Landau–Wilson functional with the quartic interaction term. We consider first the case of thermal transitions and then in the next chapter we study the quantum case. The aim is to show that the RG not only provides the concepts to understand critical phenomena but is also a powerful tool to obtain the values for the critical exponents.

We start with the expression for the Landau–Wilson functional close to the critical point:

$$F = \int d^d x \left[\frac{1}{2} |\nabla M|^2 + \frac{r}{2} M^2 + \frac{u}{4!} M^4 \right]. \tag{4.1}$$

We now write the order parameter as a sum of a uniform contribution and a spacing-varying part:

$$M(x) = M_0 + \delta M(x) \tag{4.2}$$

such that

$$M^2(x) \approx M_0^2 + \delta M^2 \tag{4.3}$$

and

$$M^4(x) \approx M_0^4 + 6M_0^2 \delta M^2. \tag{4.4}$$

Substituting in Eq. 4.1, we get up to order δM^2:

$$F = \int d^d x \left[\frac{1}{2} |\nabla \delta M(x)|^2 + \frac{r}{2}(M_0^2 + \delta M^2) + \frac{u}{4}(M_0^4 + M_0^2 \delta M^2) \right] \tag{4.5}$$

$$= \Omega \left[\frac{r}{2} M_0^2 + \frac{u}{4!} M_0^4 \right] + \int d^d x \left[\frac{1}{2} |\nabla \delta M(x)|^2 + \frac{r}{2} \delta M^2 + \frac{u}{4} M_0^2 \delta M^2 \right] \tag{4.6}$$

where $\Omega = L^d$ is the volume of the system. We can still rewrite this as

$$= \Omega \left[\frac{r}{2} M_0^2 + \frac{u}{4!} M_0^4 \right] + \int d^d x \frac{1}{2} \delta M(x) \left[-\nabla^2 + r + \frac{u}{2} M_0^2 \right] \delta M(x) \tag{4.7}$$

where $|\nabla \delta M|^2$ has been replaced by $-\delta M \nabla^2 \delta M$ by performing a partial integration and neglecting the surface term at infinity. We now divide the fluctuations of the order parameter in *fast* and *slow* fluctuations, i.e. $\delta M = \delta M_> + \delta M_<$, where the slow modes are given by

$$\delta M_<(x) = \int\limits_{0 \geq |q| < k} \frac{d^d q}{(2\pi)^d} e^{iqx} \delta M_<(q) \tag{4.8}$$

and the fast modes by

$$\delta M_>(x) = \int\limits_{k \geq |q| \leq \Lambda} \frac{d^d q}{(2\pi)^d} e^{iqx} \delta M_>(q), \tag{4.9}$$

where Λ is an ultraviolet cut-off in momentum space and $0 < k < \Lambda$ an arbitrary wave vector.

4.2 The Renormalisation Group in Momentum Space

Slow Modes

The contribution of the slow modes for the functional is given by

$$\delta F_< = \int\limits_{0 < |q| < k} d^d q \frac{1}{2} \delta M_<(q) [q^2 + r + \frac{u}{2} M_0^2] \delta M_<(-q) \tag{4.10}$$

We recall that the original integral had the limits $q \in [0, \Lambda]$. Let us restore these limits, making the following change of variables:

$$q' = \frac{\Lambda q}{k} \tag{4.11}$$

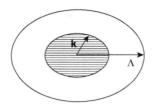

Figure 4.1 A scale transformation in momemtum space. The length scale factor $b = \Lambda/k$.

or $q' = bq$, which corresponds in real space to a scaling transformation by a factor, $b = \Lambda/k$, in the size of the system, i.e. from length scale L to $L' = L/b$.

We then get

$$\delta F_< = \left(\frac{k}{\Lambda}\right)^d \int_{0<|q'|<\Lambda} d^d q' \frac{1}{2} \delta M_<(\frac{kq'}{\Lambda}) \left[\frac{k^2}{\Lambda^2} q'^2 + r + \frac{u}{2} M_0^2\right] \delta M_<\left(-\frac{kq'}{\Lambda}\right). \tag{4.12}$$

The order parameter is rescaled as

$$\delta M'_<(q') = s^{-1} \delta M_<(\frac{kq'}{\Lambda}), \tag{4.13}$$

where s depends on the length scale factor b. Then,

$$\delta F_< = \left(\frac{k}{\Lambda}\right)^d s^2 \int_{0<|q'|<\Lambda} d^d q' \frac{1}{2} \delta M'_<(q') \left[\frac{k^2}{\Lambda^2} q'^2 + r + \frac{u}{2} M_0^2\right] \delta M'_<(-q'). \tag{4.14}$$

In order to restore the original form of the q^2 term of the functional we must have

$$\left(\frac{k}{\Lambda}\right)^d s^2 \left(\frac{k}{\Lambda}\right)^2 = 1, \tag{4.15}$$

which leads to

$$s = \left(\frac{k}{\Lambda}\right)^{-\frac{(d+2)}{2}}. \tag{4.16}$$

We then have

$$\delta F_< = \int_{0<|q'|<\Lambda} d^d q' \frac{1}{2} \delta M'_<(q') \left[q'^2 + \left(\frac{k}{\Lambda}\right)^{-2} r + \left(\frac{k}{\Lambda}\right)^{-2} \frac{u}{2} M_0^2\right] \delta M'_<(-q'). \tag{4.17}$$

Since the idea of the RG method is to obtain the rescaled functional in the same form of the original one, we define

$$r' = \left(\frac{k}{\Lambda}\right)^{-2} r. \tag{4.18}$$

In order to obtain the rescaling of the coefficient of the quartic term u, we need to obtain the renormalisation of the uniform order parameter in the new lattice. For this we write (see Eqs. 4.6 and 4.7):

$$\Omega r M_0^2 = \Omega \left(\frac{k}{\Lambda}\right)^2 r' M_0^2. \tag{4.19}$$

Since the volume scales as

$$\Omega = \left(\frac{k}{\Lambda}\right)^{-d} \Omega' \tag{4.20}$$

we find

$$\Omega r M_0^2 = \left(\frac{k}{\Lambda}\right)^{-d} \Omega' \left(\frac{k}{\Lambda}\right)^2 r' M_0^2 = \left(\frac{k}{\Lambda}\right)^{2-d} \Omega' r' M_0^2. \tag{4.21}$$

Defining M_0' through, $\Omega r M_0^2 = \Omega' r' M_0'^2$, we finally obtain

$$M_0^2 = \left(\frac{k}{\Lambda}\right)^{d-2} M_0'^2. \tag{4.22}$$

For the u-term we now get

$$\frac{u'}{2} M_0'^2 = \left(\frac{k}{\Lambda}\right)^{-2} \frac{u}{2} M_0^2 = \left(\frac{k}{\Lambda}\right)^{-2} \frac{u}{2} \left(\frac{k}{\Lambda}\right)^{d-2} M_0'^2, \tag{4.23}$$

where we used the previous equation for the renormalisation of M_0^2. We finally obtain

$$u' = \left(\frac{k}{\Lambda}\right)^{d-4} u. \tag{4.24}$$

Recognising the scaling factor $b = (\Lambda/k)$ in Eqs. 4.18 and 4.24, we obtain the RG equations which describe the change of the parameters r and u under the length scale transformation, as

$$r' = b^2 r$$
$$u' = b^{4-d} u. \tag{4.25}$$

These equations can also be written in differential form as

$$k \frac{\partial r'}{\partial} = -2r'^2 \tag{4.26}$$

$$k\frac{\partial u'}{\partial k} = (d-4)u' \tag{4.27}$$

and for the order parameter

$$k\frac{\partial M_0'^2}{\partial k} = -(2-d)M_0'^2. \tag{4.28}$$

The fixed points of the RG equations, Eqs. 4.25, are given by, $(r^*, u^*) = (0,0)$, which we can identify as corresponding to the Gaussian fixed point. Notice that the quartic interaction u is *irrelevant* for $d > 4$, i.e. it scales to zero under successive renormalisations. Consequently, the Gaussian functional, with no quartic term, and the Gaussian fixed point give the correct exponents for $d > 4$, as we obtained in the previous chapter. Starting from a distance r_0 of the critical point and iterating N times the equation for r' up to a distance $l = b^N$, we get

$$r' = l^2 r_0. \tag{4.29}$$

Since r' is arbitrary, we take $r' = 1$, to obtain $l^2 r_0 = 1$. We take this value of r' to define the scale of the correlation length, $l = \xi = r_0^{-1/2} = |T - T_c|^{-1/2}$, from which we recognise the Gaussian correlation length exponent, $\nu = 1/2$. Note, on the other hand, that for $d > 4$, $u' = l^{4-d} u_0$ vanishes when $l \to \infty$. This is just another way of expressing the irrelevance of u for $d > d_c = 4$. Alternatively, this can be seen as a determination of the upper critical dimension $d_c = 4$ since for $d > 4$, u vanishes under renormalisation and the Gaussian or mean field approach becomes exact.

Eqs. 4.22 or 4.28 are used to identify the Gaussian exponent, $\eta = 0$, of the order parameter correlation function, which is defined through the relation $M_0'^2 = b^{d-2+\eta} M_0^2$.

Integration of Fast Modes

The part of the functional associated with the fast modes is given by

$$F_> = \int_k^{\Lambda} d^d q \frac{1}{2} \delta M_>(q) \left[q^2 + r + \frac{u}{2} M_0^2 \right] \delta M_>(-q). \tag{4.30}$$

We shall now integrate the fast modes, which correspond to short wavelength fluctuations, to obtain an effective functional given by

$$e^{-\Omega F_{eff}} = \frac{e^{-\Omega F_0} \int_{-\infty}^{+\infty} \prod_{|q|>k}^{\Lambda} \delta M_>(q) \delta M_>(-q) e^{-\frac{1}{2}\int_k^{\Lambda} d^d q \delta M_>(q)[q^2+r+\frac{u}{2}M_0^2]\delta M_>(-q)}}{\int_{-\infty}^{+\infty} \prod_{|q|>k}^{\Lambda} \delta M_>(q) \delta M_>(-q) e^{-\frac{1}{2}\int_k^{\Lambda} d^d q \delta M_>(q)[q^2+r]\delta M_>(-q)}}.$$

$$\tag{4.31}$$

The integrals can be performed, changing the inner integrals into sums and using that

$$\int \prod dx_i e^{-\frac{1}{2}\sum_{i,j} A_{ij}x_i x_j} = \sqrt{\frac{\pi}{Det A}}$$ (4.32)

to get

$$\int_{-\infty}^{+\infty} \prod_{|q|>k}^{\Lambda} \delta M_>(q)\delta M_>(-q)e^{-\frac{1}{2}\int_k^{\Lambda} d^d q \delta M_>(q)[q^2+r+\frac{u}{2}M_0^2]\delta M_>(-q)}$$

$$= \prod_{|q|>k}^{\Lambda} \left[\frac{\pi}{q^2 + r + \frac{u}{2}M_0^2}\right]^{1/2}$$

$$= e^{-\frac{1}{2}\sum_{|q|>k}^{\Lambda} \ln[q^2+r+\frac{u}{2}M_0^2]+\frac{1}{2}(\sum_{|q|>k}^{\Lambda})\ln\pi}.$$ (4.33)

Then,

$$e^{-\Omega F_{eff}} = \frac{e^{-\Omega F_0}e^{-\frac{1}{2}\sum_{|q|>k}^{\Lambda} \ln[q^2+r+\frac{u}{2}M_0^2]}}{e^{-\frac{1}{2}\sum_{|q|>k}^{\Lambda} \ln[q^2+r]}}$$ (4.34)

and we finally obtain for the effective functional or *effective potential*,

$$F_{eff} = F_0 + \frac{1}{2\Omega}\sum_{|q|>k}^{\Lambda} \ln\left[q^2 + r + \frac{u}{2}M_0^2\right] - \frac{1}{2\Omega}\sum_{|q|>k}^{\Lambda} \ln\left[q^2 + r\right],$$ (4.35)

which we rewrite as

$$F_{eff} = F_0 + F_k = F_0 + \frac{K_d}{2}\int_k^{\Lambda} dq q^{d-1}\left[\ln\left[q^2 + r + \frac{u}{2}M_0^2\right] - \ln\left[q^2 + r\right]\right],$$ (4.36)

where

$$F_0 = \frac{r}{2}M_0^2 + \frac{u}{4!}M_0^4.$$ (4.37)

We have substituted, $\sum_k = \Omega/(2\pi)^3 \int d^d q$ with $\int d^d q = S_d \int dq q^{d-1}$, such that $S_d = 2\pi^{d/2}/\Gamma(d/2)$ is the area of a unit d-dimensional sphere. The quantity $K_d = S_d/(2\pi)^d = 2/(4\pi)^{d/2}\Gamma(d/2)$ with $\Gamma(x)$ the *Gamma function*.

Let us find how the effective functional changes under a change of scale

$$\frac{\partial F_{eff}}{\partial k} = \frac{\partial F_k}{\partial k} = -\frac{K_d}{2}k^{d-1}\ln\frac{k^2 + r + (u/2)M_0^2}{k^2 + r},$$ (4.38)

which can be rewritten as

$$k\frac{\partial F_k}{\partial k} = -\frac{K_d}{2}k^d \ln\left[1 + \frac{uM_0^2}{2(k^2 + r)}\right]. \tag{4.39}$$

Expanding for small u, we get

$$k\frac{\partial F_k}{\partial k} = -\frac{K_d}{2}k^d\left[\frac{uM_0^2}{2(k^2 + r)} - \frac{u^2M_0^4}{8(k^2 + r)^2}\right]. \tag{4.40}$$

Let us consider the original form of the functional F_0 but with length or k-dependent couplings, such that

$$F_0^k = \frac{r(k)}{2}M_0^2 + \frac{u(k)}{4!}M_0^4. \tag{4.41}$$

Then,

$$\frac{\partial F_0^k}{\partial k} = \frac{1}{2}\frac{\partial r(k)}{\partial k}M_0^2 + \frac{1}{24}\frac{\partial u}{\partial k}M_0^4 \tag{4.42}$$

or

$$k\frac{\partial F_0^k}{\partial k} = \frac{1}{2}k\frac{\partial r(k)}{\partial k}M_0^2 + \frac{1}{24}k\frac{\partial u}{\partial k}M_0^4. \tag{4.43}$$

Comparing the coefficients of the powers of M_0^2 and M_0^4 of this equation with those of Eq. 4.40, we find

$$k\frac{\partial r(k)}{\partial k} = -\frac{K_d}{2}k^d\frac{u}{2(k^2 + r)}$$

$$k\frac{\partial u}{\partial k} = \frac{K_d}{2}k^d\frac{u^2}{8(k^2 + r)^2}.$$

Let us define

$$r = k^2\bar{r}$$

$$u = k^{4-d}\bar{u},$$

such that

$$k\frac{\partial\bar{r}}{\partial k} = -2\bar{r} + \frac{1}{k}\frac{\partial r}{\partial k}$$

$$k\frac{\partial\bar{u}}{\partial k} = -(4 - d)\bar{u} + k^{d-3}\frac{\partial u}{\partial k}.$$

These finally yield the RG equations at the one-loop level:

$$k\frac{\partial\bar{r}}{\partial k} = -2\bar{r} - \frac{K_d\bar{u}}{2(1 + \bar{r})} \tag{4.44}$$

$$k\frac{\partial\bar{u}}{\partial k} = -(4 - d)\bar{u} + \frac{3}{2}\frac{K_d\bar{u}^2}{(1 + \bar{r})^2}. \tag{4.45}$$

4.3 Fixed Points

The equations above can be written in the alternative form:

$$\frac{\partial \bar{r}}{\partial \ln b} = 2\bar{r} + \frac{K_d \bar{u}}{2(1 + \bar{r})} \tag{4.46}$$

$$\frac{\partial \bar{u}}{\partial \ln b} = (4 - d)\bar{u} - \frac{3}{2}\frac{K_d \bar{u}^2}{(1 + \bar{r})^2} \tag{4.47}$$

where $b = \Lambda/k$. Before obtaining the fixed points we expand the above equations to $O(r^2, ur, u^2)$ to get

$$\frac{\partial \bar{r}}{\partial \ln b} \approx 2\bar{r} + \frac{K_d \bar{u}}{2}(1 - \bar{r}) = \bar{r}(2 - \frac{K_d \bar{u}}{2}) + \frac{K_d \bar{u}}{2} \tag{4.48}$$

$$\frac{\partial \bar{u}}{\partial \ln b} \approx (4 - d)\bar{u} + \frac{3}{2}K_d \bar{u}^2 = \bar{u}(\epsilon - \frac{3K_d}{2}\bar{u}) \tag{4.49}$$

with $\epsilon = 4 - d$.

The fixed points of the RG equations are the solutions of $(\partial \bar{r}/\partial \ln b) = (\partial \bar{u}/\partial \ln b) = 0$ and can be easily obtained. We get $(\bar{u}_G^* = 0, \bar{r}_G^* = 0)$, which is the Gaussian fixed point and another fixed point at $(\bar{u}^* = 2\epsilon/3K_d, \bar{r}^* = -\epsilon/(6 - \epsilon))$. Then we have found a new fixed point different from the Gaussian fixed point which, as we show below, controls the critical behaviour of the system in dimensions $(4 - \epsilon)$, smaller than the upper critical dimension, $d_c = 4$. Since we have expanded for small u (see Eq. 4.40), we expect ϵ to be small to guarantee the limit of validity of this ϵ-*expansion*.

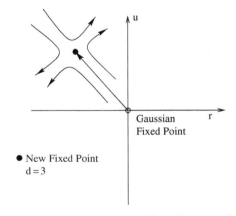

Figure 4.2 The fixed points of the Landau–Wilson functional in parameter space. The Gaussian fixed point is unstable with respect to the quartic interaction u for $d < 4$. The arrows indicate the flow of the renormalisation group equations. The full dot is the new fixed point which appears going beyond the Gaussian approximation.

Then for small ϵ, the non-trivial fixed point at $(\bar{u}^* = 2\epsilon/3K_d, \bar{r}^* = -\epsilon/[6 - \epsilon])$ is located at a small value of $\bar{u}$ and this justifies the expansion of the RG equations for small values of this quartic interaction. Even if the actual physical value of the quartic term $\bar{u}$ is large, the flow of the renormalisation group equations, as shown in Fig. 3.2 for $d < 4$, is towards the non-trivial fixed point at small $\bar{u}$.

4.4 Renormalisation Group Flows and Critical Exponents

In order to study the flow of the renormalisation group equations in the parameter space (r, u), it is useful to work with a discretised form of the RG equations. In this way the iteration of the equations leads to a vivid visualisation of the flows and the relevant or irrelevant character of the different variables. We start integrating the linearised RG equations, Eqs. 4.49, to obtain

$$\bar{r}' - \bar{r} = \left[2\bar{r} - \bar{r}\frac{K_d\bar{u}}{2} + \frac{K_d\bar{u}}{2} \right] \ln b \qquad (4.50)$$

$$\bar{u}' - \bar{u} = \left[\epsilon\bar{u} - \frac{3K_d}{2}\bar{u}^2 \right] \ln b \qquad (4.51)$$

with $b \geq 1$. Using the relation $b^x - 1 \approx x \ln x$, we get

$$\bar{r}' = b^2\bar{r} + \bar{u}(b^{K_d/2} - 1) - \bar{r}\bar{u}(b^{K_d/2} - 1)$$
$$\bar{u}' = b^\epsilon\bar{u} - \bar{u}^2(b^{3K_d/2} - 1). \qquad (4.52)$$

It is easy to check that in the limit $b \to 1$, the fixed points of these equations, $\bar{r}^* = 0, \bar{u}^* = 0$ and $(\bar{r}^* = \bar{u}^*/(\bar{u}^* - (b^2 - 1)/(b^{K_d/2} - 1)), \bar{u}^* = (b^\epsilon - 1)/(b^{3K_d/2} - 1))$ coincide with those obtained previously. These equations can now be easily iterated. Starting from different pairs $(\bar{r}_0, \bar{u}_0)$ and $b = 1.1$, for example, we use the recursion relations above to find the image of this point under the transformations, Eqs. 4.52, the image of this image and so on, to obtain the (schematic) flow diagram shown in Fig. 3.2.

In the neighbourhood of each fixed point, the RG equations can be expanded in the form

$$\begin{pmatrix} \Delta\bar{r}' \\ \Delta\bar{u}' \end{pmatrix} = \begin{pmatrix} \frac{\partial\bar{r}'}{\partial\bar{r}} & \frac{\partial\bar{r}'}{\partial\bar{u}} \\ \frac{\partial\bar{u}'}{\partial\bar{r}} & \frac{\partial\bar{u}'}{\partial\bar{u}} \end{pmatrix} \begin{pmatrix} \Delta\bar{r} \\ \Delta\bar{u} \end{pmatrix},$$

where the Jacobian is calculated at the fixed point. The eigenvalues at the Gaussian fixed point are

$$\lambda_1 = b^2$$
$$\lambda_2 = b^\epsilon.$$

At the non-trivial fixed point they are given by

$$\lambda_1 = b^2 - \overline{u}^*(b^{K_d/2} - 1)$$
$$\lambda_2 = b^\epsilon - 2\overline{u}^*(b^{3K_d/2} - 1).$$

The scaling fields, i.e. the eigenvectors, are linear combinations in the form $(\overline{r} - \overline{r}^*) + w(\overline{u} - \overline{u}^*)$, with w determined from the eigenvalue equations. In practice, we can write

$$\Delta\overline{r}' = b^{\lambda_1}\Delta\overline{r}$$
$$\Delta\overline{u}' = b^{\lambda_2}\Delta\overline{u}$$

with $\Delta\overline{r} \propto (T - T_c)$ and $\Delta\overline{u} = (\overline{u} - \overline{u}^*)$.

The critical exponents can now be obtained. For the correlation length exponent we get

$$\nu^{-1} = \left(\frac{\partial\lambda_1}{\partial\ln b}\right)_{b=1} = 2 - \frac{\epsilon}{3},$$

such that, at the Gaussian fixed point ($u = 0$), $\nu_G^{-1} = 1/2$. On the other hand, at the non-trivial fixed point, $(\partial\lambda_2/\partial\ln b)_{b=1} = -\epsilon$. This negative quantity, for $d < 4$, is not related to any critical exponent but gives rise only to scaling corrections. The important physical significance of this quantity is that it implies that the flow of the quartic interaction, for $d < 4$, is towards the non-trivial fixed point at $\overline{u}^*$. Note that the equivalent quantity at the Gaussian fixed point is given by, $\left(\partial\lambda_2^G/\partial\ln b\right)_{b=1} = \epsilon$, which is positive for $d < 4$, implying that the quartic interaction is a relevant field at the Gaussian fixed point for $d < 4$. When ϵ changes sign, there is an exchange of stability between the Gaussian and the non-trivial fixed points, such that, for $d > 4$ the critical behaviour is controlled by the former. At $d = 4$, or $\epsilon = 0$, both fixed points merge and this characterises the marginal behaviour at the upper critical dimension. The critical behaviour in this case is mean field-like, with logarithmic corrections (Toulouse and Pfeuty, 1975).

The table below gives the values of the critical exponents in $3d$ for a one-component magnetic (Ising) system calculated with the simple first-order ϵ-expansion presented above. We used that the exponent $\eta = \eta_G + O(\epsilon^2)$, with the Gaussian value $\eta_G = 0$ obtained previously, and the scaling relations $\gamma = (2 - \eta)\nu$ and $2\beta = (d - 2 + \eta)$ to get the critical exponents β and γ,

	Theory $O(\epsilon)$	High-T Expansion
ν	0.58333	0.6333
γ	1.1667	1.24
β	0.33333	0.33

This represents a considerable improvement with respect to the mean field exponents when compared to the high-temperature expansion results, which can be taken as exact values (Cardy, 1996). Much better agreement however can still be obtained going to higher orders in the ϵ-expansion (Cardy, 1996).

4.5 Conclusions

In this chapter, we have presented an implementation of the momentum space renormalisation group that allows the critical exponents of the finite-temperature magnetic phase transition of an Ising system in three dimensions to be obtained. We used a first-order approach in an ϵ-expansion that already gives rise to a non-trivial fixed point at which we can obtain non-Gaussian exponents. Although in $3d$, $\epsilon = 4 - d = 1$ is not a small quantity, the momentum space RG gives consistently better results for the critical exponents at higher orders in ϵ (Brezin, 2014). We used the properties of the effective potential under length scale transformations to obtain the RG equations. This quantity when minimised yields the ground state of the system. The renormalisation group, which is based on the idea of length scale transformations, associates a mathematical entity, the fixed point and the notion of scale invariance with a critical point of an actual physical system undergoing a phase transition.

5

Quantum Phase Transitions

5.1 Effective Action for a Nearly Ferromagnetic Metal

In this chapter we use the renormalisation group to study quantum phase transitions (Hertz, 1976). As before, we present the method by applying it to a specific problem, in this case the zero-temperature transition from a paramagnetic to a ferromagnetic metal. For quantum transitions, the *effective action* replaces the Landau–Wilson functional as the basic quantity containing the essential physics of the problem. Instead of starting with the effective action for the quantum paramagnetic-to-ferromagnetic transition in a metal, we consider first the microscopic Hamiltonian from which it is derived. This is the Hubbard model (Hubbard, 1963) for a nearly ferromagnetic metal, which is given by

$$H = H_0 + H_1 = \sum_{k,\sigma} \epsilon_k c_{k\sigma}^\dagger c_{k\sigma} + U \sum_i n_i^\uparrow n_i^\downarrow, \tag{5.1}$$

where the first term H_0 describes the kinetic energy of the electrons with wave vector k and spin σ. The second term H_1 takes into account the Coulomb repulsion between electrons with opposite spins in the same site i. It can be rewritten as a sum of charge and spin terms in the following way:

$$U \sum_i n_i^\uparrow n_i^\downarrow = \frac{U}{4} \sum_i (n_i^\uparrow + n_i^\downarrow)^2 - \frac{U}{4} \sum_i (n_i^\uparrow - n_i^\downarrow)^2.$$

The statistical properties of the system are obtained from its partition function. Furthermore, because we are interested in nearly magnetic systems, we neglect the charge term and keep only the magnetic contribution proportional to $(S^z)^2 = (\sum_\sigma \sigma n_{i\sigma})^2$ in the Hamiltonian. Using the operator identity,

$$e^{-\beta H} = e^{-\beta H_0} T \exp\left(-\int_0^\beta d\tau \, H_1(\tau)\right),$$

57

where $H_1(\tau) = e^{\tau H_0} H_1 e^{-\tau H_0}$, T is the time ordering operator and $\beta = 1/k_B T$, we can write

$$e^{-\beta H} = e^{-\beta H_0} T \exp\left(-\frac{U}{4} \sum_i \int_0^\beta d\tau [\sum_\sigma \sigma n_{i\sigma}(\tau)]^2\right) \tag{5.2}$$

$$= e^{-\beta H_0} T \exp\left(-\sum_i \int_0^\beta d\tau \frac{[\sum_\sigma \sigma \sqrt{\frac{U}{2}} n_{i\sigma}(\tau)]^2}{2}\right). \tag{5.3}$$

We can remove the power of the term in the time-ordered exponential using the Hubbard–Stratanovich relation:

$$e^{\frac{a^2}{2}} = \int_{-\infty}^{+\infty} \frac{dx}{\sqrt{2\pi}} e^{-\frac{x^2}{2} - ax}.$$

Also using that $\int_0^\beta f(\tau) d\tau = \lim_{M\to\infty} \frac{\beta}{M} \sum_{n=1}^M f(\tau_n)$, where $\tau_n = \beta(n/M)$, we can write the partition function of the system as

$$Z = Tr e^{-\beta H} = Z_0 \int D\psi e^{-\frac{1}{2} \sum_i \int_0^\beta \psi_i^2(\tau) d\tau} \left\langle T e^{-\sum_{i,\sigma} \int_0^\beta d\tau \sigma V_i(\tau) n_{i\sigma}(\tau)} \right\rangle_0, \tag{5.4}$$

where $D\psi = (2\pi)^{N/2} \prod_i d\psi_i(\tau)$ and $V_i(\tau) = \sqrt{U/2}\psi_i(\tau)$ is an effective time-dependent magnetic field acting in the magnetic moment at site i. The average of the operator is $< A >_0 = (1/Z_0) Tr e^{-\beta H_0} A$, with $Z_0 = Tr e^{-\beta H_0}$. Introducing the non-interacting Green's function, associated with H_0,

$$G_{ij}^0(\tau, \tau') = \left\langle T c_{i\sigma}(\tau) c_{j\sigma}^\dagger(\tau') \right\rangle = \frac{1}{\beta} \sum_{k,n} \frac{e^{ik.(R_i - R_j) - iE_n(\tau - \tau')}}{iE_n - \epsilon_k}$$

we can rewrite the partition function as

$$Z = Z_0 \int D\psi \exp\left(-\frac{1}{2} \sum_i \int_0^\beta \psi_i^2(\tau) d\tau + \sum_\sigma Tr \ln(1 - \sigma V G^0)\right). \tag{5.5}$$

This expression emphasises that only one-particle Green's functions appear in the expansion of the exponential, since H_0 is quadratic in fermion operators. The matrix $V_{ij}(\tau, \tau') = V_i(\tau) \delta_{ij} \delta(\tau - \tau')$. From the expansion

$$Tr \ln(1 - \sigma V G^0) = -\sum_{n=1}^\infty \frac{1}{2n} Tr(V G^0)^{2n}$$

we see that only even powers of V appear in the expansion of the exponential. Let us consider the second-order term of this expansion. This is given by

$$Tr[(V G_0)^2] = \sum_{il} \int\int d\tau d\tau'' V_i(\tau) V_l(\tau'') G_{il}^0(\tau, \tau'') G_{li}^0(\tau'', \tau). \tag{5.6}$$

Introducing Fourier transforms such that

$$V_i(\tau) = \sum_{k,n} V_k(\omega_n) e^{ik.r_i + i\omega_n \tau}$$

and

$$G_{ij}^0(\tau, \tau') = \sum_{kn} G_k^0(\omega_n) e^{ik.(r_i - r_j) + i\omega_n(\tau - \tau')},$$

we get

$$Tr[(VG_0)^2] = \sum_{k_1 k_3 \omega_1 \omega_3} V_{k_1}(\omega_1) V_{-k_1}(-\omega_1) G_{k_3}^0(\omega_3) G_{k_1+k_3}^0(\omega_1 + \omega_3)$$

$$= -\sum_{k_1 \omega_1} |V_{k_1}(\omega_1)|^2 \chi_0(k_1, \omega_1),$$

where

$$\chi_0(q, \omega) = -\frac{1}{\beta} \sum_{k_3 \omega_3} G_{k_3}^0(\omega_3) G_{q+k_3}^0(\omega + \omega_3),$$

which can be written as

$$\chi_0(q, \omega) = \sum_q \frac{f(\epsilon_{k+q}) - f(\epsilon_k)}{\epsilon_k - \epsilon_{k+q} + \omega},$$

where $f(\epsilon)$ is the Fermi function and ϵ_k the dispersion relation of the electrons. For the free electron gas, this is the Lindhard function, given by

$$\chi_0(q, \omega) = N(\epsilon_F)[1 + \frac{\omega}{2v_F q} \ln \frac{\omega - v_F q}{\omega + v_F q}],$$

where $N(\epsilon_F)$ is the density of states at the Fermi level and $v_F = q_F/m$ is the Fermi velocity. For small q/q_F and $\omega/v_F q_F$ this is expanded as

$$\chi_0(q, \omega) \approx N(\epsilon_F)[1 + i\frac{\pi \omega}{2v_F q} - \frac{q^2}{12q_F^2} + \cdots]. \tag{5.7}$$

The term on the top of the exponential in the equation for the partition function, see Eqs. 5.4 and 5.5, can be viewed as a functional $\Phi(\psi)$ of the field $\psi_i(\tau)$, which couples to the magnetisation of the original Hubbard model. Since it acts as a pseudo-magnetic field on the local magnetisation of the system, a finite ψ implies a finite magnetisation and consequently it is acceptable to take this field as the order parameter. This functional, or *effective action*, $\Phi(\psi)$ can be written as

$$\Phi(\psi) = \frac{1}{2} \sum_{q\omega} v_2(q, \omega) |\psi(q, \omega)|^2 + \frac{1}{4\beta} \sum_{q_i \omega_i} v_4(q_1, ..., q_4, \omega_1, ..., \omega_4) \psi(q_1, \omega_1)$$

$$\times \psi(q_2, \omega_2) \psi(q_3, \omega_3) \psi(q_4, \omega_4) \delta(\sum_{i=1}^{4} q_i) \delta(\sum_{i=1}^{4} \omega_i) + \cdots$$

$$+ \frac{1}{m(\beta N)^{m/2-1}} \sum_{q_i \omega_i} v_m(q_1,.....q_m, \omega_1, ...\omega_m)$$

$$\times \prod_{i=1}^{m} \psi(q_i, \omega_i)\delta(\sum_{i=1}^{m} q_i)\delta(\sum_{i=1}^{m} \omega_i) + \cdots$$

where $v_2(q, \omega) = 1 - U\chi_0(q, \omega)$. In the quantum case the relevant fields are functions of frequency as well as of momentum. This frequency dependence ultimately arises due to the non-commutativity of the competing terms in the Hamiltonian.

5.2 The Quantum Paramagnetic-to-Ferromagnetic Transition

At zero temperature, the action for a nearly ferromagnetic system in Euclidean space of dimension d is expanded close to the quantum critical point (QCP) as

$$F = \frac{1}{2} \int d^d q \int d\omega v_2(q, \omega)|\psi(q, \omega)|^2 + \frac{u}{4} \int d^d q_1 \cdots \int d^d q_4 \int d\omega_1 \cdots$$

$$\times \int d\omega_4 \psi(q_1, \omega_1)\psi(q_2, \omega_2)\psi(q_3, \omega_3)\psi(q_4, \omega_4), \qquad (5.8)$$

where $\sum_i \vec{q}_i = 0$ and $\sum_i \omega_i = 0$. The integrals over the components of the wave vector $\vec{q}$ are done over the first Brillouin zone with a cut-off $\Lambda = (\pi/a)$. We now separate the field ψ in a uniform static and a dynamic, wave vector-dependent contribution:

$$\psi(q, \omega) = \psi_0 \delta(\omega)\delta_{q,0} + \delta\psi(q, \omega). \qquad (5.9)$$

As in the classical case, we take for the ψ^4 term the following decoupling:

$$\psi(q_1, \omega_1)\psi(q_2, \omega_2)\psi(q_3, \omega_3)\psi(q_4, \omega_4) \approx \psi_0^4 + 6\psi_0^2 \delta\psi(q, \omega)\delta\psi(-q, -\omega) \qquad (5.10)$$

such that

$$F \approx L^{(d+z)} F_0 + \int d^d q \int d\omega \frac{1}{2}\delta\psi(q, \omega)\left[v_2(q, \omega) + 3u\psi_0^2\right]\delta\psi(-q, -\omega), \qquad (5.11)$$

where L is the size of the system and the dynamic exponent z takes into account the extra dimensions associated with the *time directions*. Also

$$F_0 = \left[\frac{g}{2}\psi_0^2 + \frac{u}{4}\psi_0^4\right],$$

where

$$g = \lim_{q \to 0} \lim_{\omega \to 0} v_2(q, \omega) = 1 - UN(\epsilon_F)$$

measures essentially the distance to the quantum critical point (QCP). This can be written as $g = |(U/W) - (U/W)_c|$, where $W = 1/N(\epsilon_F)$ is the bandwidth and $(U/W)_c$ the critical ratio for the appearance of ferromagnetism. For the nearly ferromagnetic metal, $v_2(q, \omega) = 1 - U\chi_0(q, \omega)$ is obtained using the small wave vector, low frequency expansion of the Lindhard function and is given by

$$v_2(q, \omega) = g + q^2 + \frac{|\omega|}{q},$$

where we have changed units in order to simplify the action.

Slow Modes

Let us consider the contribution for the last term of the effective action arising from the slow modes, i.e. from modes such that $0 < |q| < k$. Again, as in the thermal case of the previous chapter, we perform a change of variables to bring the limit of the integrals back to the original one. For that we need the following change of variables:

$$q' = \frac{\Lambda}{k}q$$
$$\omega' = \left(\frac{\Lambda}{k}\right)^z \omega.$$

The dynamic exponent z is introduced above as the exponent which rescales frequency under a length scale by a factor $b = \Lambda/k$. With these transformations we obtain:

$$F_< = \left(\frac{k}{\Lambda}\right)^d \left(\frac{k}{\Lambda}\right)^z \int_0^\Lambda d^d q' \int d\omega' \frac{1}{2} \left|\delta\psi[\frac{k}{\Lambda}q', (\frac{k}{\Lambda})^z\omega']\right|^2 [\frac{k^2}{\Lambda^2}q'^2 + g + \frac{(k/\Lambda)^z\omega'}{(k/\Lambda)q'} + 3u\psi_0^2].$$

(5.12)

The field scales according to

$$\delta\psi'(q', \omega') = s^{-1}\delta\psi[\frac{k}{\Lambda}q', (\frac{k}{\Lambda})^z\omega'].$$

In order to keep the first term (arising from the gradient), fixed, we impose

$$\left(\frac{k}{\Lambda}\right)^d \left(\frac{k}{\Lambda}\right)^z \left(\frac{k}{\Lambda}\right)^2 s^2 = 1$$

such that

$$s = \left(\frac{k}{\Lambda}\right)^{-\frac{d+z+2}{2}}$$

and

$$F_< = \int_0^\Lambda d^d q' \int d\omega' \frac{1}{2} |\delta\psi'(q', \omega')|^2 [q'^2 + (\frac{k}{\Lambda})^{-2}g$$
$$+ \frac{(k/\Lambda)^{-2}(k/\Lambda)^z\omega'}{(k/\Lambda)q'} + 3u(\frac{k}{\Lambda})^{-2}\psi_0^2].$$

(5.13)

We now get

$$g' = b^2 g, \tag{5.14}$$

where the scaling factor $b = (\Lambda/k)$. The quantity ω'/q' is chosen to remain unrenormalised, which implies that

$$\frac{(k/\Lambda)^{-2}(k/\Lambda)^z}{(k/\Lambda)} = 1.$$

This equation determines the value for the dynamic exponent z of a nearly ferromagnetic metal, which is given by

$$z = 3.$$

The scaling of the uniform, static part of the order parameter is obtained by considering its contribution for f. We get

$$L^{d+z} g \psi_0^2 = (bL')^{d+z} b^{-2} g' \psi_0^2 \equiv (L')^{d+z} g' \psi_0'^2.$$

The last identity defines the scaled uniform static field

$$\psi_0'^2 = b^{d+z-2} \psi_0^2. \tag{5.15}$$

Note that since the scaling dimension of the magnetisation correlation function is that of ψ_0^2, we can obtain from the equation above the Gaussian exponent, $\eta = 0$, since this is defined through, $\psi_0'^2 = b^{d+z-2+\eta} \psi_0^2$.

For the renormalisation of the coefficient of the ψ^4 term we get

$$u' = b^{4-(d+z)} u. \tag{5.16}$$

These equations, together with Eq. 5.14, are the renormalisation group equations for the quantum case obtained from the slow modes. The fixed point is at $(g^* = 0, u^* = 0)$, which we immediately recognise as the Gaussian fixed point. We note in the last two equations that the Euclidean dimension appears in the combination $d + z$, such that the effective dimension is in fact $d_{eff} = d + z$. Extremely important is the fact that the scaling dimension of the quartic interaction u, according to the equation above, is $4 - (d + z)$. Then for $d_{eff} = d + z > 4$ this is an *irrelevant* perturbation at the Gaussian fixed point. This implies that in this case, the quantum critical behaviour is described by Gaussian exponents which are all known. There are of course many important and relevant physical problems for which the condition $d + z > 4$ is satisfied; hence the importance of this result.

Integration of Fast Modes

In this case the effective action is obtained from

$$
e^{\Omega F_{eff}} = \frac{e^{-L^{d+z}F_0}\int\limits_{|q|>k,\omega>k^z}^{\Lambda,\Lambda^z}\prod \delta\psi(q,\omega)\delta\psi(-q,-\omega)e^{-\frac{1}{2}\int_k^\Lambda d^dq\int_{k^z}^{\Lambda^z}d\omega[g+q^2+\frac{\omega}{q}+3u\psi_0^2]|\delta\psi(q,\omega)|^2}}{\int\limits_{|q|>k,\omega>k^z}^{\Lambda,\Lambda^z}\prod \delta\psi(q,\omega)\delta\psi(-q,-\omega)e^{-\frac{1}{2}\int_k^\Lambda d^dq\int_{k^z}^{\Lambda^z}d\omega[g+q^2+\frac{\omega}{q}]|\delta\psi(q,\omega)|^2}}.
$$

The integration over the fast modes can be easily performed, like in the classical case, and we get

$$
e^{\Omega F_{eff}} = \frac{e^{-L^{d+z}F_0}e^{-\frac{1}{2}\sum'\ln[g+q^2+\frac{\omega}{q}+3u\psi_0^2]}}{e^{-\frac{1}{2}\sum'\ln[g+q^2+\frac{\omega}{q}]}}.
$$

Consequently, we can write for the effective action

$$
F_{eff} = F_0 + \frac{1}{2L^{d+z}}\left[\sum{}'\ln[g+q^2+\frac{\omega}{q}+3u\psi_0^2] - \sum{}'\ln[g+q^2+\frac{\omega}{q}]\right]
$$

or

$$
F_{eff} = F_0 + \frac{1}{2L^{d+z}}\sum{}'\ln[1+\frac{3u\psi_0^2}{\left(g+q^2+\frac{\omega}{q}\right)}]. \tag{5.17}
$$

Expanding the ln to second power in u, we get

$$
F_{eff} = F_0 + \frac{1}{2L^{d+z}}\left[\sum{}'\frac{3u}{\left(g+q^2+\frac{\omega}{q}\right)}\psi_0^2 - \sum{}'\frac{9u^2}{2\left(g+q^2+\frac{\omega}{q}\right)^2}\psi_0^4\right].
$$

Now let us examine carefully the meaning of the sum $\sum' = \sum'_{q,\omega}$. This represents, in fact, an integration over wave vectors and frequencies such that

$$
\frac{1}{L^{d+z}}\sum_{q,\omega}{}' = \frac{2S_d}{(2\pi)^{d+1}}\left[\int_{k^z}^{\Lambda^z}d\omega\int_0^\Lambda q^{d-1}dq + \int_0^{\Lambda^z}d\omega\int_k^\Lambda q^{d-1}dq\right].
$$

These are carried out over an infinitesimal L-shaped region, as shown in Fig. 5.1. The factor 2 comes from the frequency integration over positive and negative values. The quantity S_d is the area of a d-dimensional unit sphere. We now define $C_d = 2S_d/(2\pi)^{d+1}$ and write the effective action as

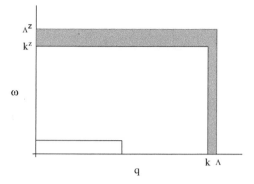

Figure 5.1 The region of integration for the quantum renormalisation group.

$$F_{eff} = F_0 + \frac{3C_d u \psi_0^2}{2} \left[\int_{k^z}^{\Lambda^z} d\omega \int_0^{\Lambda} q^{d-1} dq \frac{1}{g+q^2+\frac{\omega}{q}} + \int_0^{\Lambda^z} d\omega \int_k^{\Lambda} q^{d-1} dq \frac{1}{g+q^2+\frac{\omega}{q}} \right]$$

$$- \frac{9C_d u^2 \psi_0^4}{4} \left[\int_{k^z}^{\Lambda^z} d\omega \int_0^{\Lambda} q^{d-1} dq \frac{1}{\left(g+q^2+\frac{\omega}{q}\right)^2} + \int_0^{\Lambda^z} d\omega \int_k^{\Lambda} q^{d-1} dq \frac{1}{\left(g+q^2+\frac{\omega}{q}\right)^2} \right].$$

Let us take $\Lambda = 1$ and calculate the quantity

$$\frac{\partial F_{eff}}{\partial k} = -\frac{3C_d u \psi_0^2}{2} \left[zk^{z-1} \int_0^1 dq \frac{q^{d-1}}{g+q^2+\frac{k^z}{q}} + k^{d-1} \int_0^1 d\omega \frac{1}{g+k^2+\frac{\omega}{k}} \right]$$

$$+ \frac{9C_d u^2 \psi_0^4}{4} \left[zk^{z-1} \int_0^1 dq \frac{q^{d-1}}{\left(g+q^2+\frac{k^z}{q}\right)^2} + k^{d-1} \int_0^1 d\omega \frac{1}{\left(g+k^2+\frac{\omega}{k}\right)^2} \right]$$

where we used that $\frac{\partial}{\partial k} = zk^{z-1} \frac{\partial}{\partial k^z}$.

Now we define

$$\overline{g} = g/k^2$$
$$\overline{q} = q/k$$
$$\overline{\omega} = \omega/k^z = \omega/k^3$$
$$\overline{u} = u/k^{4-(d+z)} = u/k^{1-d}$$

to obtain

$$\frac{\partial F_{eff}}{\partial k} = -\frac{3C_d u k^d}{2}\left[z\int_0^{1/k}\overline{q}^d d\overline{q}\frac{1}{g\overline{q}+\overline{q}^3+1} + \int_0^{1/k^z}d\overline{\omega}\frac{1}{\overline{g}+1+\overline{\omega}}\right]\psi_0^2$$

$$+\frac{9C_d u^2 k^{d-2}}{4}\left[z\int_0^{1/k}d\overline{q}\frac{\overline{q}^{d+1}}{\left(g\overline{q}+\overline{q}^3+1\right)^2} + \int_0^{1/k^z}d\overline{\omega}\frac{1}{(\overline{g}+1+\overline{\omega})^2}\right]\psi_0^4.$$

The new variables obey the equations

$$k\frac{\partial\overline{g}}{\partial k} = -2\overline{g}+\frac{1}{k}\frac{\partial g}{\partial k}$$

$$k\frac{\partial\overline{u}}{\partial k} = -[4-(d+z)]\overline{u}+k^{d+z-3} = -[4-(d+z)]\overline{u}+k^d\frac{\partial u}{\partial k}.$$

As in the previous chapter, we consider the uniform, static action but with k-dependent coefficients

$$F_0^k = \frac{g(k)}{2}\psi_0^2 + \frac{u(k)}{4}\psi_0^4 \tag{5.18}$$

such that

$$\frac{\partial F_0^k}{\partial k} = \frac{1}{2}\frac{\partial g(k)}{\partial k}\psi_0^2 + \frac{1}{4}\frac{\partial u(k)}{\partial k}\psi_0^4. \tag{5.19}$$

Comparing the terms of the same order in ψ_0 of this equation with those of Eq. 5.18, we obtain the following renormalisation group equations:

$$k\frac{\partial\overline{g}}{\partial k} = -2\overline{g}-3C_d\overline{u}\left[z\int_0^1 d\overline{q}\frac{\overline{q}^d}{g\overline{q}+\overline{q}^3+1} + \int_0^1 d\overline{\omega}\frac{1}{\overline{g}+1+\overline{\omega}}\right]$$

$$k\frac{\partial\overline{u}}{\partial k} = -[4-(d+z)]\overline{u}+9C_d\overline{u}^2\left[z\int_0^1 d\overline{q}\frac{\overline{q}^{d+1}}{\left(g\overline{q}+\overline{q}^3+1\right)^2} + \int_0^1 d\overline{\omega}\frac{1}{(\overline{g}+1+\overline{\omega})^2}\right]$$

for k close to $\Lambda = 1$ ($b \approx 1$). This can still be written in a more usual form (Hertz, 1976) as

$$\frac{\partial\overline{g}}{\partial\ln b} = 2\overline{g}+3C_d\overline{u}\left[z\int_0^1 d\overline{q}\frac{\overline{q}^d}{g\overline{q}+\overline{q}^3+1} + \int_0^1 d\overline{\omega}\frac{1}{\overline{g}+1+\overline{\omega}}\right] \tag{5.20}$$

$$\frac{\partial\overline{u}}{\partial\ln b} = [4-(d+z)]\overline{u}-9C_d\overline{u}^2\left[z\int_0^1 d\overline{q}\frac{\overline{q}^{d+1}}{\left(g\overline{q}+\overline{q}^3+1\right)^2} + \int_0^1 d\overline{\omega}\frac{1}{(\overline{g}+1+\overline{\omega})^2}\right].$$

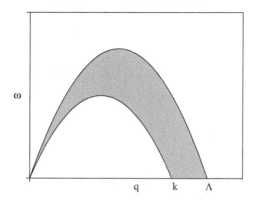

Figure 5.2 The simplified cut-off. The removed degrees of freedom correspond
to the shaded area between the two curves.

A Simplified Cut-off

The renormalisation group equations can be obtained in a more simple form if,
instead of using independent frequency and momentum cut-offs, as in Fig. 5.1,
we use a simplified cut-off, where the frequency cut-off is taken as being momentum dependent. Let us remove modes, such that the frequency and momentum
integrations are limited to the range,

$$k^2 < \frac{|\omega|}{q} + q^2 < \Lambda^2,$$

which corresponds to integrating out the modes in the region shown in Fig. 5.2,
between the curves $\omega_1(q) = k^2 q - q^3$ and $\omega_2(q) = \Lambda^2 q - q^3$. The relevant
frequency and momentum integrations is

$$I = C_d \int_0^\Lambda dq \, q^{d-1} \int_{\omega_1}^{\omega_2} d\omega = \frac{C_d \Lambda^{d+3}}{d+1}(1 - e^{-2\ln b}),$$

where, as before, $b = (\Lambda/k)$. For $k \approx \Lambda = 1$, the quantity $(|\omega|/q + q^2) = 1$,
for the modes being integrated out. In the limit $b \to 1$, the renormalisation group
recursion relations assume in this case a much simpler form, given by,

$$\frac{\partial \bar{g}}{\partial \ln b} = 2\bar{g} + \frac{3\tilde{K}_d \bar{u}}{\bar{g} + 1} \tag{5.21}$$

$$\frac{\partial \bar{u}}{\partial \ln b} = [4 - (d+z)]\bar{u} - \frac{9\tilde{K}_d \bar{u}^2}{(\bar{g}+1)^2},$$

which resemble the form obtained in the case of thermal phase transitions. The
quantity $\tilde{K}_d = 2C_d/(d+1)$.

The equations above have two fixed points: the Gaussian one at $(\overline{g}^* = 0, \overline{u}^* = 0)$ and a non-trivial fixed point. The Gaussian fixed point is stable for $d + z > 4$, in the sense that, whenever this condition is satisfied, the quartic interaction is an irrelevant perturbation at the Gaussian fixed point. Consequently for $d + z > 4$ the quantum critical behaviour is governed by Gaussian exponents. This is the case of the nearly ferromagnetic system investigated here, for which in Euclidean dimension ($d = 3$), the simple Gaussian fixed point controls the zero temperature paramagnetic-to-ferromagnetic transition and describes correctly the quantum critical behaviour of this system.

5.3 Extension to Finite Temperatures

The renormalisation group equations obtained above are valid at zero temperature where frequency is a continuum variable. We would like now to extend this approach to finite but small temperatures where we may expect quantum fluctuations should manifest in the thermal behaviour of real physical systems. As we saw in Chapter 1, close to the zero-temperature unstable fixed point, temperature behaves as an *effective field* with a scaling dimension given by the dynamic exponent z. The continuous version of the RG equation for the temperature close to the quantum critical point is given by:

$$\frac{dT}{dl} = zT(l).$$

We will be interested in the case $d + z > 4$, where the relevant unstable fixed point is Gaussian. Expanding the renormalisation group equations in the neighbourhood of this fixed point, we get

$$\frac{d\overline{g}}{dl} = \frac{1}{\nu}\overline{g}(l) + \overline{u}\,f[T(l)]$$
$$\frac{d\overline{u}}{dl} = -\theta\overline{u}(l),$$

where $\theta = (d + z) - 4$ and $\nu = 1/2$. This a generalisation of the quantum RG equations, Eqs. 5.21, for finite temperatures, which are obtained by substituting the frequency integrals by sums over Matsubara frequencies. This is the origin of the term, $\overline{u}\,f[T(l)]$, in the first equation and which couples the quartic interaction $\overline{u}$ to temperature. It arises from Eq. 5.20 for small $\overline{u}$ and finite T. The system above can be solved perturbatively in the small parameter $\overline{u}$. First we integrate the equations for T and $\overline{u}$, to obtain

$$T = T_0 e^{z(l-l_0)}$$
$$\overline{u} = \overline{u}_0 e^{-\theta(l-l_0)},$$

which, when substituted in the Eq. 5.22 for $\overline{g}(l)$, yields

$$\frac{d\overline{g}(l)}{dl} = \frac{\overline{g}(l)}{v} + \overline{u}_0 e^{-\theta(l-l_0)} f[T_0 e^{z(l-l_0)}],$$

which can be rewritten as

$$\frac{d\overline{g}(l)}{\overline{g}(l)} = \frac{dl}{v} + \frac{\overline{u}_0 e^{-\theta l}}{\overline{g}(l)} f[T_0 e^{zl}] dl,$$

where we took $l_0 = 0$. Since we are interested in a solution to first order in u, we take the zero-order solution for $\overline{g}$ in this coupling, namely

$$\overline{g}(l) = \overline{g}_0 e^{\frac{l}{v}},$$

and substitute it in the RG equation for $\overline{g}(l)$ above to get

$$\frac{d\overline{g}(l)}{\overline{g}(l)} = \frac{dl}{v} + \frac{\overline{u}_0 e^{-(\theta+1/v)l}}{\overline{g}_0} f[T_0 e^{zl}] dl.$$

Integrating from 0 to l and expanding to first order in $\overline{u}_0$, we get

$$\overline{g}(l) = \overline{g}_0 e^{\frac{l}{v}} + \overline{u}_0 e^{\frac{l}{v}} \int_0^l dl' e^{-(\theta+1/v)l'} f[T_0 e^{zl'}].$$

Changing variables, this equation becomes

$$\overline{g}(l) = \overline{g}_0 e^{\frac{l}{v}} + \frac{\overline{u}_0 e^{\frac{l}{v}} T^{\frac{(\theta+1/v)}{z}}}{z} \int_T^{T e^{lz}} \frac{dx}{x} x^{-(\theta+1/v)/z} f(x)$$

and since $L = e^l$, we obtain

$$\overline{g}(b) = L^{1/v}[\overline{g}_0 + \frac{\overline{u}_0 T^{\frac{1}{\psi}}}{z} \int_T^{T L^z} \frac{dx}{x} x^{-1/\psi} f(x)],$$

where $\psi^{-1} = \frac{(\theta+1/v)}{z}$. The correlation length $\xi(T)$ is defined as the length scale at which $\overline{g}(L) = 1$, or $L = \xi$, such that

$$1 = \xi^{1/v}(T)[\overline{g}_0 + \frac{\overline{u}_0 T^{\frac{1}{\psi}}}{z} \int_T^{T \xi^z} \frac{dx}{x} x^{-1/\psi} f(x)]. \tag{5.22}$$

Now for $T \to 0$, but $T\xi^z \to \infty$, we take the integral as a constant and find for the correlation length

$$\xi = \frac{1}{|\overline{g}_0 + \overline{u}_0 T^{\frac{1}{\psi}}|^v}, \tag{5.23}$$

where unimportant constants have been made equal to one. Then the correlation length diverges at a temperature dependent critical line given by

$$T_c \propto \left(\frac{|\overline{g}_0|}{\overline{u}_0}\right)^{\psi}. \tag{5.24}$$

The exponent, ψ which characterises the behaviour of the line of finite temperature phase transitions near the quantum critical point, is the *shift exponent*, introduced in Chapter 1, and in the present case it is given by

$$\psi^{-1} = \frac{(\theta + 1/\nu)}{z} = \frac{d + z - 2}{z}. \tag{5.25}$$

This result was obtained by Millis (Millis, 1993). It relates the shift exponent to exponents associated with the zero-temperature fixed point. It implies a break-down of the generalised scaling relation $\psi = \nu z$ due to dangerously irrelevant interaction u.

Notice that the shape of the critical line can be measured close to the quantum critical point and yields information about the zero-temperature critical exponents.

The function $f(x)$ in Eq. 5.22 is such that $f(T \to 0) \propto T^2$ and $f(T \to \infty) \propto T$ (Sachdev *et al.*, 1995). For the limits we assumed for the integral in this equation to hold, the shift exponent must satisfy the following inequalities, $1/2 < \psi < 1$, such that, for the critical line, $(dT_c/d\,\overline{g}_0)_{\overline{g}_0=0} = \infty$. For $\psi \leq 1/2$, the shift exponent is fixed at the mean field value obtained in Chapter 1, $\psi^{-1} = 2$.

In the treatment above, the correlation length diverges at the critical line with the same Gaussian exponent, $\nu = 1/2$, of the zero-temperature transition. However, while for $d + z > 4$, the Gaussian exponents are exact for the zero-temperature phase transition, *this is not the case* for the finite-temperature phase transitions. Sufficiently close to the critical line, the critical behaviour is governed by *Wilson exponents,* different from those associated with the zero-temperature fixed point as we saw in Chapter 1.

Although the quartic interaction $\overline{u}_0$ is *irrelevant* at the Gaussian fixed point, for $d + z > 4$, it plays a crucial role in determining the equation for the critical line T_c, as shown by Eq. 5.24. Also, the behaviour along the critical trajectory, $\overline{g}_0 = 0$, $T \to 0$, depends crucially on $\overline{u}_0$. For example, the correlation length along this trajectory is given by (see Eq. 5.23)

$$\xi(\overline{g}_0 = 0, T) = \frac{1}{\overline{u}_0^\nu T^{\frac{\nu}{\psi}}} \tag{5.26}$$

with $\overline{u}_0$ appearing in the denominator, which emphasises the *dangerously irrelevant* character of this quartic interaction for $d + z > 4$, since it cannot just be taken as being equal to zero.

In Section 1.7, we presented a general scaling analysis of the critical behaviour of a quantum phase diagram with a temperature axis. We introduced a set of *tilde* (Wilson) exponents, different from the zero-temperature critical exponents, which characterise the critical behaviour along the critical line T_c. We obtained for the correlation length at the quantum critical point:

$$\xi(\overline{g}_0 = 0, T) \propto T^{\frac{\tilde{\nu}-\nu}{\nu z}}(uT^{\frac{1}{\psi}})^{-\tilde{\nu}}.$$

A comparison between this general result and the previous expression, Eq. 5.26, for the correlation length shows that the former arises, taking $\nu = \tilde{\nu}$ in the equation above. Then the result, Eq. 5.26, relies on the identity between the thermal and quantum correlation length exponents, i,e., $\tilde{\nu} = \nu$. Note also that if $\psi = \nu z$, we recover the result of naive scaling of Chapter 1.

5.4 Effective Action Close to a Spin-Density Wave Instability

In this section we discuss the effective action for a metallic system close to a spin-density wave instability characterised by a wave vector Q ($Q \neq 0$). We shall study situations where there is no nesting. In this case, following similar methods to those used in the ferromagnetic case, we find that the zero-temperature effective action can be written as

$$F_{eff} = \int d^d q \int d\omega \frac{1}{2}[g + q^2 + \frac{|\omega|}{\tau}]|\psi(q,\omega)|^2 + \frac{u}{4} \int \prod_{i=1}^{4} d^d q_i$$

$$\times \int \prod_{i=1}^{4} d\omega_i \psi(q_1,\omega_1)\psi(q_2,\omega_2)\psi(q_3,\omega_3)\psi(q_4,\omega_4),$$

(5.27)

where $\sum_i \vec{q}_i = 0$, $\sum_i \omega_i = 0$. The quantity g measures the distance to the quantum critical point and τ is the q-independent lifetime of the spin fluctuations (*paramagnons*) with wave vector $Q \pm q$. The field ψ is the *order parameter*. We consider systems above their upper critical dimension and present magnetic order for some range of the parameters at finite temperatures. Differently from the action of the nearly ferromagnetic metal, Eq. 5.8, the damping τ^{-1} of the paramagnons is now q-independent. The reason is that in the ferromagnet, the total magnetisation is a conserved quantity and has to diffuse over large distances, in correspondingly long times. For a spin-density wave instability of wave vector $Q \neq 0$, the order parameter is not a conserved quantity and the paramagnons may decay locally (Sachdev, 1999). This difference has profound implications which are essentially contained in the different values for the dynamic exponents z. The rescaling transformations performed earlier for the nearly ferromagnetic metal may be easily repeated here to yield the dynamic exponent $z = 2$, distinct from the value $z = 3$ of the former case. The action, Eq. 5.27 describes the important case of a nearly antiferromagnetic metal which is relevant for high-T_c superconductors (Millis, Monien and Pines, 1990) and heavy fermions (Millis, 1993).

When we carry out the RG calculations for the finite Q instability, the basic results we obtain are the scaling expression for the quartic interaction u:

$$u' = b^{4-(d+z)}u, \qquad (5.28)$$

and the value of the dynamic exponent $z = 2$. These equations tell us that the upper critical dimension $d_c = 2$, such that, for $d \geq 2$, the Gaussian fixed point is stable against the interaction between the fluctuations and the quantum ($T = 0$) phase transition from the paramagnetic metal to the Q-ordered spin-density wave state is governed by Gaussian exponents.

5.5 Gaussian Effective Actions and Magnetic Instabilities in Metallic Systems

Since for quantum phase transitions the effective dimension is $d + z$, for magnetic transitions in metallic systems with $z = 3$ or $z = 2$, Gaussian effective actions should play an important role, as they yield the exact critical exponents whenever $d + z > d_c = 4$. Consequently, we devote this section to a closer look into these actions. We consider the finite-temperature Gaussian effective action

$$F_{eff}^G = \frac{1}{\beta} \sum_{\omega_n} \int \frac{d^d q}{(2\pi)^d} \frac{1}{2} \psi(q, \omega_n) \left[g + q^2 + \tau(q)|\omega_n| \right] \psi(-q, -\omega_n), \qquad (5.29)$$

where $\tau(q) = 1/(\Gamma q)$ for the nearly ferromagnetic metal and $\tau = 1/\Gamma$ for an incipient spin-density wave state. The free energy f_G is obtained after integrating over the ψ-fields

$$e^{-\beta f_G} = \int \prod_{q,\omega_n} d\psi(q, \omega_n) d\psi(-q, -\omega_n) e^{-F_{eff}^G[\psi(q,\omega_n)]} \qquad (5.30)$$

to yield

$$f_G = \sum_{\omega_n} \int \frac{d^d q}{(2\pi)^d} \ln \left[g + q^2 + \tau(q)|\omega_n| \right]. \qquad (5.31)$$

The sum over bose frequencies can be performed using the Kramers–Kronig relation:

$$\begin{aligned} \ln M(\omega_n) &= \frac{1}{\pi} \int dx \frac{\Im m \ln M^R(x)}{-i\omega_n + x} \\ &= \frac{1}{\pi} \int dx \frac{\tan^{-1}[\Im m M^R(x)/\Re e M^R(x)]}{-i\omega_n + x}, \end{aligned} \qquad (5.32)$$

where $M^R(x)$ is the argument of the logarithm written in the real frequency axis. The frequency sum can now be performed and we obtain for the Gaussian free energy density of a metal close to a spin-density wave instability:

$$f_G = \frac{1}{2\pi} \int_{-\infty}^{+\infty} d\omega \coth(\beta\omega/2) \int_0^\Lambda \frac{d^d q}{(2\pi)^d} \tan^{-1}\left[\frac{\tau(q)\omega}{g + q^2}\right].$$

In the nearly ferromagnetic case, $\tau(q) \propto 1/q$ and the dynamic exponent $z = 3$. In the other cases τ is a constant and $z = 2$. The results below are common to all cases, independent of the wavevector Q of the magnetic instability:

- The free energy density, for $\Lambda\xi \to \infty$ where the correlation length $\xi = |g|^{-\nu}$ with $\nu = 1/2$, can be cast in the scaling form

$$f_G \propto |g|^{\nu(d+z)} F[\frac{T}{T_{coh}}], \tag{5.33}$$

  where the *coherence line* $T_{coh} = |g|^{\nu z}$. The quantity g measures the distance to the quantum critical point.
- For temperatures below the coherence line, $T \ll T_{coh}$, the system behaves as a *Fermi liquid* in the sense that the free energy density has an expansion in even powers of T, i.e. the scaling function in the equation above behaves as

$$F[\frac{T}{T_{coh}} \ll 1] \approx 1 + \left(\frac{T}{T_{coh}}\right)^2 + \left(\frac{T}{T_{coh}}\right)^4 + \cdots \tag{5.34}$$

- The specific heat in the Fermi liquid regime, for $T \ll T_{coh}$, is given by

$$C/T \propto \gamma_0 + c|g|^{\nu(d-z)}, \tag{5.35}$$

  where c is a constant. The second term defines the scaling contribution to the *thermal mass*, $m_T = C/T$, of the Fermi liquid. Whenever $d = z$, this term is replaced by $\ln g$.
- At the quantum critical point, i.e. at $g = 0$, the specific heat, for example, behaves as

$$C/T \propto T^{\frac{d-z}{z}}, \tag{5.36}$$

  which, in general, is *non-Fermi liquid*-like. Again, for $d = z$, the term on the right-hand side is replaced by $\ln T$.

These results are obtained at the Gaussian level and do not take into account the effect of the dangerously irrelevant quartic interaction u.

5.6 Field-Dependent Free Energy

Let us now consider the nearly magnetic system in the presence of an external field conjugate to the order parameter. In the case of the nearly ferromagnetic metal this is a uniform magnetic field and a *staggered field* in the antiferromagnetic case. The scaling form of the free energy density near a quantum critical point in the presence of an external field h conjugate to the order parameter was shown in Chapter 1 to be given by

$$f_S \propto |g|^{2-\alpha} F_s \left[\frac{T}{T_{coh}}, \frac{h}{|g|^{\beta+\gamma}} \right]. \tag{5.37}$$

The order parameter $m \propto (\partial f_s / \partial h)_{h=0} \propto |g|^{\beta}$ and the susceptibility $\chi \propto (\partial^2 f_S / \partial h^2)_{h=0} \propto |g|^{-\gamma}$. The exponents α, β and γ obey the usual scaling relations derived in Chapter 1. In particular, the quantum hyperscaling relation $2 - \alpha = \nu(d + z)$ is satisfied for $d + z < d_c$ with d_c the upper critical dimension. In the case of the ferromagnet where the order parameter, the total magnetisation, is a conserved quantity, Sachdev (Sachdev, 1994) has shown that the conjugate (uniform) magnetic field scales close to the QCP as

$$h' = b^z h \tag{5.38}$$

which implies the additional relation

$$\beta + \gamma = \nu z. \tag{5.39}$$

Although Eq. 5.39 is strictly valid for an insulator, it is also satisfied for the nearly ferromagnetic metal.

We now make an important distinction between Gaussian quantum theories and the mean field one. Let us consider the Gaussian action studied above at $T = 0$. Since in this case the susceptibility exponent $\gamma_G = 1$ and $\alpha_G = 2 - 1/2(d + z)$, in order to satisfy the scaling relations, $\alpha + 2\beta + \gamma = 2$ and $\gamma = \beta(\delta - 1)$, we must have, $\beta_G = \frac{1}{4}(d + z - 2)$ and $\delta_G = (d + z + 2)/(d + z - 2)$. Together with $\nu_G = 1/2$ these are the quantum Gaussian exponents. These values are clearly different from the mean field ones, $\beta = 1/2$ and $\delta = 3$ for all $d + z > d_c = 4$. As can be easily verified, for $d + z > 4$, the Gaussian exponents β_G and δ_G above yield a less singular or sub-dominant behaviour for the order parameter when compared with the mean field results. This can be traced back to the observation that the Gaussian exponent, $\alpha_G = 2 - \frac{1}{2}(d + z)$, becomes negative for $d + z > 4$, while the mean field α remains fixed at $\alpha_{mf} = 0$ for all $d + z \geq 4$. A coherent picture of these results within the scaling theory is obtained considering the scaling form of the zero-temperature free energy density in the presence of the quartic interaction u

$$f_S \propto |g|^{\nu(d+z)} F_u[u|g|^{(d+z-4)/2}]$$

and assuming that u behaves as a *dangerously irrelevant* variable for $d + z > 4$, i.e. the scaling function $F_u(x \to 0) \propto 1/x$. *Irrelevant* stands for the fact that for $d + z > 4$ the term $u|g|^{(d+z-4)/2}$ approaches zero as $g \to 0$. *Dangerously* because of the non-analytic dependence of the scaling function F_u when $u \to 0$. Then, near the quantum critical point we get

$$f_S \propto \frac{|g|^{\nu(d+z)}}{u|g|^{(d+z-4)/2}} \propto \frac{|g|^2}{u}.$$

This is indeed the form of the free energy in the long-range ordered magnetic phase, within the mean field approximation, as we have shown in Chapter 2 where the quartic interaction u appears in the denominator of the free energy expression (see Eq. 2.2). The quantum case is obtained just by replacing $(T - T_c)$ by g. The critical exponent α is now fixed at the mean field value $\alpha = 0$ for all $d+z > 4$. This is another instance of breakdown of a scaling relation, in this case of hyperscaling since the relation $2 - \alpha = \nu(d + z)$ is now violated for $d + z > 4$. Previously, in Section 5.3, we obtained the failure of the relation $\psi = \nu z$, also due to the dangerously irrelevant interaction u.

In conclusion, taking into account the effect of the quartic interaction, at $T = 0$, we are able to reconcile the Gaussian theory with the mean field results which actually describe the quantum critical behaviour at zero temperature and in the presence of a field conjugate to the order parameter. The approach presented in Section 5.3 does not yield the corrections above. There, u appears coupled to temperature and at $T = 0$ it just does not play any role.

5.7 Gaussian versus Mean Field at $T \neq 0$

Let us discuss further some subtle aspects of the physics above the upper critical dimension at finite but small temperatures ($T \ll T_{coh}$). The total free energy in the non-critical side of the phase diagram, below the coherence line, can be written as a sum of an analytic and a Gaussian contribution,

$$f_t = |g|^2 + |g|^{\nu(d+z)} F[\frac{T}{|g|^{\nu z}}],$$

where we neglected the temperature dependence of the analytic contribution assuming it is less singular than that of the Gaussian. In the Fermi liquid regime, i.e. $T \ll T_{coh} = |g|^{\nu z}$, we have

$$f_t = |g|^2 + |g|^{\nu(d+z)} \left\{ 1 + \left(\frac{T}{|g|^{\nu z}} \right)^2 + \cdots \right\}.$$

Then, sufficiently close to the critical point and for $d + z > 4$, with $\nu = 1/2$, we obtain

$$f_t = |g|^2 + |g|^{\nu(d+z)} \left(\frac{T}{|g|^{\nu z}} \right)^2 + \cdots .$$

We neglected the zero-temperature Gaussian term since $\nu(d+z) > 2$ for $d+z > 4$, and this is less singular than the analytic term when $g \to 0$. The free energy density can be rewritten in the scaling form:

$$f_t = |g|^2 F[\frac{T}{T_{sf}}].$$

The new *spin-fluctuation temperature* (Beal-Monod *et al.*, 1968) is given by

$$T_{sf} = |g|^{1 - \frac{d-z}{4}},$$

where we used $\nu = 1/2$. Consequently, the effect of the analytic contribution is to introduce a new energy scale in the Fermi liquid region of the phase diagram, namely, T_{sf}. If we calculate the specific heat from the above expression for the free energy, we get $C/T \propto |g|^{\nu(d-z)}$, which of course coincides with the Gaussian result. The new energy scale plays a role when we take into account the field dependence of the analytic (mean field) and Gaussian contributions. Then, it can easily be shown that the order parameter susceptibility, $\chi_0 = \left(\partial^2 f / \partial h^2 \right)_{h=0}$, with

$$f \propto |g|^2 F_0(T/T_{sf}, h/|g|^{\beta+\gamma})$$

is given by $\chi_0 = |g|^{-1} F_1(T/T_{sf})$ for $T \ll T_{coh}$, where we used the mean field exponents, $\beta = 1/2$ and $\gamma = 1$. The field h here is that conjugated to the order parameter.

We point out that, for $d + z > 4$, even at $|g| = 0$ there is a characteristic field h_{cross} at which there is a crossover from mean field to Gaussian behaviour. The order parameter m at the quantum critical point varies with the conjugate field h, according to $m \propto h^{1/\delta}$. For small fields, i.e. $h \ll h_{cross}$, the mean field contribution with $\delta_{MF} = 3$ dominates, while in the opposite case the Gaussian one with $\delta_G = (d + z + 2)/(d + z - 2)$ is dominant. This is the case, for example, at $d = 3$ and $z = 2$, where $\delta_G = 7/3 < \delta_{MF} = 3$.

5.8 Critique of Hertz Approach

The concept of effective dimension and its implications for quantum phase transitions may be used with care for systems at and below the lower critical dimension. Systems at these dimensionalities have strong thermal and quantum fluctuations, and it is not even clear if they present a long-range ordered ground state at zero temperature. It is now widely accepted that Hertz theory, that integrates out the fermion modes, fails for $d \leq 2$. The self-energy of these fermions has an imaginary part proportional to their energy to a power that is equal or less than one, implying that they are not well-defined excitations at least in some regions of the Fermi surface. Removing these modes in these conditions does not lead to a trustful action of the type we have considered. We will not discuss these points further, here, since they are extensively discussed in Sachdev's book (Sachdev, 2011).

Another possible source of difficulties for Hertz description of a magnetic quantum phase transition in metals is the coupling of the bosonic modes associated with the magnetic order parameter to low frequency particle-hole excitations in the metal. This coupling may change the nature of the phase transition at zero temperature from second to first order. We will have the opportunity to study this type of effect further on in this book.

6

Heavy Fermions

6.1 Introduction

The theory of quantum criticality has made enormous progress in the last two decades. The reason is that many physical systems have been discovered which are close to, or can be driven to the proximity of, a quantum critical point. The nature of these QCP is diverse: they can be associated with magnetic transitions, with localisation phenomena, or with superconductivity, to name just a few cases treated in this book. The most common driving mechanisms are external pressure, physical or chemical, external magnetic fields and disorder. In spite of the variety of QCP, with fluctuations of the most different characters, we expect to find some universal behaviour and a unifying perspective for these phenomena. The scaling approach has sufficiently embracing concepts as to provide this type of description, but a deep understanding of quantum critical systems demands that it comes accompanied by microscopic models. These models are necessary to determine the critical exponents and fully characterise the different universality classes.

In this chapter we add some substance to the ideas presented so far, examining a class of materials that historically has played an important role in the study of quantum criticality. Heavy fermions are found by chance to be close to some instability such that small changes in external parameters can give rise to a new phase. This can be approached sufficiently smoothly so as to appear that the phase transition is continuous, of the second order. The theory of quantum criticality provides an elegant framework to understand many of the properties of these systems, as we show in this chapter.

Heavy fermion physics is associated with elements with unstable f-shells like Ce, U and Yb (Steglich *et al.*, 1995, 1998; Kambe, Flouquet and Hargreaves, 1997; Thompson and Lawrence, 1994). The f-states in these materials have an ambiguous character, at the borderline between localised and itinerant. They form compounds with a variety of ground states, such as, magnetic (generally antiferromagnetic), superconductor or simply Fermi liquid (FL) down to the lowest temperatures. By Fermi liquids we mean here metallic systems which, below a characteristic temperature T^*, exhibit a constant, i.e. temperature-independent, uniform susceptibility χ (Pauli susceptibility), a linear temperature-dependent specific heat $C(T)$ and a resistivity ρ with a T^2 temperature dependence. The localised nature of the f-electrons involved in the heavy fermion phenomenon, implies the existence of strong correlations in these materials. However, in spite of the strong electron–electron interactions, it is found experimentally that these systems still exhibit simple Fermi liquid behaviour, i.e.

$$
\begin{aligned}
\chi(T \to 0) &= \chi_0 \\
C/T &= m_T \\
\rho &= \rho_0 + AT^2
\end{aligned}
\tag{6.1}
$$

for $T \ll T^*$, but with the parameters χ_0, m_T and A strongly renormalised by the interactions. The *thermal mass* m_T of the quasi-particles, for example, is typically two or even three orders of magnitude larger in heavy fermions than in a simple metal like copper. The characteristic temperature T^* is itself much smaller than the Fermi liquid temperature of normal metals. Notice that, from this point of view, heavy fermions are simple when compared, for example, with high-temperature superconductors which deviate radically from Fermi liquid behaviour in a wide region of parameters.

The physics of heavy fermions is governed by the competition between the Kondo effect, which gives rise to screened local moments, and the RKKY interaction, which favours the appearance of a magnetic ground state (Doniach, 1977). This competition is described by the Kondo lattice Hamiltonian, which is given by

$$
H = \sum_{k,\sigma} \epsilon_k c_{k\sigma}^\dagger c_{k\sigma} + J \sum_i \mathbf{S}_i . \sigma,
\tag{6.2.}
$$

where the first term describes a band of conduction electrons of spin σ and width W and the second the antiferromagnetic interaction between the magnetic moments of these electrons and those of the localised f-electrons, $\mathbf{S}_i$, displayed on the sites i of a regular lattice. This Hamiltonian neglects charge fluctuations, but, since the system is always metallic and we will be interested in magnetic instabilities, this is not crucial. The Hamiltonian, Eq. 6.2 has been investigated by different techniques. The general picture which arises in three dimensions is that, at zero temperature below a critical value of the ratio between the coupling J and the bandwidth W, i.e. for $(J/W) < (J/W)_c$, the RKKY interaction prevails and the ground state of the system is magnetic. For $(J/W) > (J/W)_c$ on the other hand, the Kondo screening is dominant and the ground state is non-magnetic. The phase transition at $T = 0$, $(J/W) = (J/W)_c$ is a magnetic transition from an antiferromagnetic ground state with a staggered magnetisation $m_s \neq 0$, to a non-magnetic state with $m_s = 0$. This transition is of second order and the zero-temperature order parameter m_s vanishes continuously at the quantum critical point as $m_s \propto |g|^\beta$, where $g = (J/W) - (J/W)_c$ measures the distance to the QCP and β is a critical exponent. In the critical side of the phase diagram, i.e. for $(J/W) < (J/W)_c$ and finite temperatures there is a line of second-order phase transitions which associates a Néel temperature with a given value of the ratio (J/W). This line vanishes at the QCP as $T_N = |g|^\psi$ which defines the shift exponent ψ.

The existence of a quantum critical point at $(J/W)_c$ allows us to construct a scaling theory which describes the properties of heavy fermions in the vicinity of this zero temperature phase transition (Continentino *et al.*, 1989; Continentino, 1993). Although we have used the Kondo lattice model to motivate the scaling theory presented below, the theory is more general in the sense that it relies only on the existence of the magnetic quantum critical point.

The scaling theory can be summarised in the form of the free energy density close to the QCP,

$$ f \propto |g|^{2-\alpha} F\left[\frac{T}{T_{coh}}, \frac{H}{|g|^{\beta+\gamma}}, \frac{h}{h_c} \right], \tag{6.3} $$

where $F[x, y, z]$ is a scaling function which depends on the temperature T, the staggered magnetic field H conjugated to the order parameter m_s, the uniform external field h and g, the distance to the QCP. The *coherence temperature*, $T_{coh} = |g|^{\nu z}$ is the crossover temperature in the non-critical side of the phase diagram, introduced in Chapter 1 and shown in Fig. 6.1. The characteristic uniform field, $h_c = |g|^{\phi_h}$ where ϕ_h is in principle an independent critical exponent. The exponents α, β, γ, ν and z are the zero-temperature critical exponents introduced previously. They obey scaling relations as $\alpha + 2\beta + \gamma = 2$. In particular the dynamic critical exponent z, the correlation length exponent ν and the dimensionality d of the system are related by the quantum hyperscaling relation $2 - \alpha = \nu(d + z)$ (see Eq. 1.18). For consistency we require the scaling function $F[0, 0, 0] = constant$ so that the exponent α is properly defined. On the other hand, the known experimental behaviour of heavy fermions can lead us to guess the behaviour of the scaling function in other limits.

The nature of the characteristic temperature T^* below which heavy fermion systems show Fermi liquid behaviour has been much debated. This temperature was initially identified with the Kondo temperature $k_B T_K = W e^{-1/(J/W)}$, of the impurity problem or related to it. This temperature scale however is a local property and does not incorporate the essential physics of the Kondo lattice, which is the competition between the Kondo and RKKY interactions. Continentino, Japiassu and

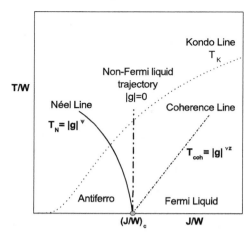

Figure 6.1 Schematic phase diagram of heavy fermions as described by the Kondo lattice model. The coherence line governed by the exponent vz and below which the system enters the Fermi liquid regime, the critical Néel line controlled by the shift exponents ψ, and the non-Fermi liquid trajectory are all shown. The quantity $g = (J/W) - (J/W)_c$ measures the distance to the QCP. Also shown is the Kondo line of the impurity problem.

Troper (1989) have proposed, on the basis of the proximity of heavy fermions to a magnetic quantum critical point, to make the following identification:

$$T^* = T_{coh} = |g|^{vz}. \tag{6.4}$$

Retrospectively, this looks natural since the existence of this QCP is the most important result of the Kondo lattice model. It is a direct consequence of the competition between Kondo and RKKY interactions and the coherence temperature T_{coh} is the new energy scale associated with the QCP. In a temperature versus (J/W) phase diagram, $T_{coh} = |g|^{vz}$ is a line in the non-critical side of the phase diagram rising from the QCP.

The identification of the characteristic energy T^*, below which the heavy fermion system enters the Fermi liquid regime, with T_{coh}, i.e., $T^* = T_{coh}$, is full of physical consequences. Let us examine the case, $H = h = 0$, such that the singular part of the free energy density can be written as

$$f_s \propto |g|^{2-\alpha} F(\frac{T}{T_{coh}}).$$

Since $T^* = T_{coh}$ marks the onset of the Fermi liquid regime, this implies that for $(T/T_{coh}) << 1$, i.e. $T << T_{coh}$, the scaling function $F(x) \approx 1 + x^2 + x^4 + \cdots$, has an expansion in even powers of $x = (T/T_{coh})$ as appropriate to a Fermi liquid. Consequently,

$$f_s(T << T_{coh}) \approx |g|^{2-\alpha}[1 + (\frac{T}{T_{coh}})^2 + (\frac{T}{T_{coh}})^4 + \cdots].$$

This Sommerfeld expansion allows the specific heat in the Fermi liquid regime to be obtained, below the *coherence line*, i.e. for $T << T_{coh}$. Since $C/T \propto \partial^2 f_s/\partial T^2$, we get to lowest order in temperature

$$C/T \approx |g|^{2-\alpha-2vz} = |g|^{v(d-z)} \tag{6.5}$$

where the last identity has been obtained using the quantum hyperscaling relation. Note that for $z \geq d$, the thermal mass of the quasi-particles $m_T \propto C/T$ is enhanced as the system gets close to the quantum phase transition (for $z = d$ this equation leads to a $\ln|g|$ dependence) from the disordered phase and below the coherence line. Naturally, this raises the expectation that we can attribute the heavy mass of heavy fermion systems to such effect. In fact, the actual explanation is different, as we shall show in the next chapter.

For the uniform susceptibility in the Fermi liquid regime we require only the free energy at $T = 0$. From $f_s(T = 0) \propto |g|^{2-\alpha} F(h/h_c)$ and $\chi_0 \propto (\partial^2 f_s/\partial h^2)_{h=0}$ we obtain $\chi_0 \propto |g|^{2-\alpha-2\phi_h}$ which for $2 - \alpha - 2\phi_h < 0$, leads to an enhanced susceptibility as the system approaches the QCP. For the resistivity, since it is not a thermodynamic quantity, we have to rely on a microscopic theory to obtain its behaviour. In any case, just based on what happens in the one-impurity Kondo problem, we expect that the same characteristic temperature which shows up in the thermodynamic quantities sets the scale for the temperature dependence of the resistivity in the Fermi liquid regime. In particular, for the dominant low-temperature contribution we expect, $\rho = \rho_0 + a(T/T_{coh})^2$, such that, in Eq. 6.2, the coefficient $A \propto T_{coh}^{-2}$.

Note that the one-impurity Kondo temperature T_K still plays a role in heavy fermion systems in the incoherent high temperature regime ($T >> T_{coh}$). Typically, in this regime, the resistivity rises with temperature, on cooling, with a logarithmic temperature dependence which is the signature of the Kondo effect. The incoherent scattering of the conduction electrons, with a short mean-free path, by the f-moments gives rise to such effect. On further cooling, the coherent regime sets in and the resistivity drops, eventually vanishing for a perfect lattice. The scale of the low temperature T^2 term in this coherent Fermi liquid regime is the characteristic lattice temperature, namely T_{coh}.

In practice, the external control parameter that allows g to be varied and the phase diagram explored is pressure. In Ce-based compounds, pressure increases the ratio (J/W) *moving* the system to the right in the phase diagram of Fig. 6.1. The experimental qualitative behaviour is that of Fig. 6.1. Starting from a magnetic ground state, pressure reduces the Néel temperature, driving the system to the QCP. In the non-critical side of the phase diagram, pressure increases the range of the

Fermi liquid regime in qualitative accord with a coherence line increasing from the QCP. An important consequence of the physical meaning of the coherence line, as the temperature below which the systems enters the Fermi liquid regime, is that at the QCP the system never enters this regime and consequently presents non-Fermi liquid behaviour. We know from the scaling theory that along the quantum critical trajectory the thermodynamic properties are governed by power law behaviour in temperature determined by critical exponents. It is not our intention either to review experimental work in heavy fermions or to make a detailed comparison between these results and the theoretical predictions for the exponents obtained from different theoretical models. It suffices here to point out that the scaling approach provides the correct qualitative picture of the phase diagram of heavy fermions. Together with microscopic models they provide the appropriate tools for a deep understanding of these materials.

In the process of applying pressure in magnetic Ce-based heavy fermions to drive it to a QCP, many experiments have confirmed the existence of a superconducting dome above the antiferromagnetic QCP. Recent work has shown that superconductivity also occurs for some Yb-based materials when antiferromagnetism is suppressed or partially suppressed. These discoveries are exciting and challenging, as superconductivity appears in a region of the phase diagram where the system presents non-Fermi liquid behaviour.

6.2 Scaling Analysis

A deep insight into the physics of heavy fermions can be gained assuming that they are close to an antiferromagnetic quantum critical point (AFQCP) and are described by a scaling theory (Continentino, 1993). The scaling approach establishes relations among the relevant critical exponents, guides us how to collapse different sets of data but does not predict in every case the values of exponents. In order to obtain the universality class of the QCP controlling the physics of these materials we need to investigate a microscopic theory such as the Kondo lattice or the nearly antiferromagnetic Fermi liquid as will be discussed in the next chapter. It is clear that the determination of the dynamical exponent z is crucial.

In this section we discuss a procedure to obtain the critical exponents of the AFQCP of different heavy fermions and establish relations between them. We will be probing the quantum critical point from the non-magnetic side of the phase diagram and below or at the coherence line. There are of course other ways to probe the AFQCP even more directly, as along the critical, *non-Fermi liquid* trajectory, $|g| = 0$, $T \to 0$ (Continentino, 1991; Löhneysen *et al.*, 1994). It is also possible to probe the quantum critical behaviour from the magnetic ordered side (Medeiros *et al.*, 2000).

For the purpose of extracting the critical exponents, we need to be able *to move* in the phase diagram of Fig. 6.1 in the neighbourhood of the QCP. In heavy fermions this is generally done through the dependence of the ratio (J/W) on volume V and consequently on pressure P, either physical or chemical. Let us define the critical volume V_c as the volume at which $(J/W) = (J/W)_c$. Consider a physical quantity which close to V_c behaves as $X(P) = A|(V - V_c)/V_c|^{-x}$ where V is the volume at pressure P. If we introduce a reference pressure P_0 (volume V_0) we have

$$\frac{X(P)}{X(P_0)} = \left(\frac{V - V_c}{V_0 - V_c}\right)^{-x} = \left(\frac{V - V_0 + V_0 - V_c}{V_0 - V_c}\right)^{-x}. \tag{6.6}$$

Defining, $\delta_0 = V_0 - V_c$ and $\Delta V = V - V_0$, where the latter gives the change in volume due to the pressure change $\Delta P = P - P_0$, we get, for ΔV sufficiently small or small changes of pressure from the reference pressure,

$$\ln\left[\frac{X(P)}{X(P_0)}\right] \approx x\kappa_0(V_0/\delta_0)\Delta P, \tag{6.7}$$

where $(V_0/\delta_0) = V_0/(V_0 - V_c) = \alpha_V/(\alpha_V - 1)$ with $\alpha_V = V_0/V_c$. For Ce compounds on the non-magnetic side of the phase diagram, $V_0 < V_c$ and $\alpha_V < 1$. The compressibility κ_0 is given by $\kappa_0 = (-1/V)(\partial V/\partial P)$. Note that in Eq. 6.7 the coefficient of ΔP depends on the critical exponent x associated with the physical quantity X. The validity of Eq. 6.7 for several physical quantities at and below the coherence line, i.e. in the Fermi liquid regime, has been verified for the heavy fermions $CeRu_2Si_2$, $CeAl_3$, UPt_3 and $CeCu_6$ as shown in Fig. 6.2 for the first three of these systems (Continentino, 1993). These materials are located in the non-critical side ($V < V_c$ or $(J/W) > (J/W)_c$) of the phase diagram of Fig. 6.1 (Doniach, 1977). For $CeAl_3$ a reference pressure of 1.2 kbars has been used to guarantee that this is the case ($CeAl_3$ is an antiferromagnet with a Néel temperature smaller than 0.5K); otherwise $P_0 = 0$.

The data plotted in this graph for different pressures are the ratios

- $m_T(P_0)/m_T(P)$ of the thermal mass, $m_T = C/T \propto \partial^2 f/\partial T^2 \propto |g|^{2-\alpha-2\nu z}$.
- $\chi(P_0)/\chi(P)$ of the limiting Pauli-like, uniform, zero-field susceptibility, $\chi_0 = \left(\partial^2 f(T = 0)/\partial^2 h\right)_{h=0} \propto |g|^{2-\alpha-2\phi_h}$.
- $T_{coh}(P)/T_{coh}(P_0)$ the coherence temperature, $T_{coh} = |g|^{\nu z}$, obtained from the maxima of $\chi_0(T)$.
- $h_c(P)/h_c(P_0)$ of the pseudo-metamagnetic field, $h_c = |g|^{\phi_h}$, obtained from the maxima of the differential susceptibility $\chi_h(h) = \partial^2 f/\partial^2 h$ for a fixed temperature $T \ll T_{coh}$.
- $[A(P_0)/A(P)]^{1/2}$, the square root of the coefficient of the T^2 term of the low temperature resistivity.

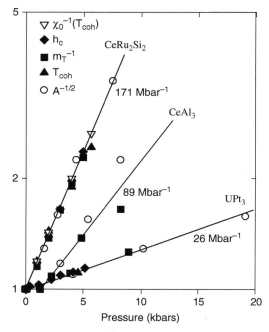

Figure 6.2 Semi-logarithmic plot $X(P)/X(P_0)$ for several physical quantities X, at or below T_{coh}, as a function of pressure for different heavy fermions. For $CeAl_3$, $P_0 = 1.2$ kbars otherwise $P_0 = 0$ (ambient pressure). The numbers close to the lines are their inclinations Γ_V that are given in Table I (see also text). Reprinted figure with permission from Continentino (1993), *Physical Review B47*, 11587. Copyright 1993 by the American Physical Society.

The collapse of the different data, for a given material, on a single line as shown in Fig. 6.2 implies that the exponent x is the same for all these quantities. It can be easily checked that this leads to the following relations among the critical exponents:

$$2 - \alpha = \nu z \qquad (6.8)$$

$$\phi_h = \nu z. \qquad (6.9)$$

Furthermore, it implies that the scale for the T^2 term of the resistivity is set by T_{coh} such that, in the expression for the low-temperature resistivity, $\rho = \rho_0 + AT^2$, we have $A \propto T_{coh}^{-2}$.

The meaning and implications of the above relations among the exponents will be discussed further down. The inclination of the lines in Fig. 6.2, i.e. $\Gamma_V = \ln\left[\frac{X(P)}{X(P_0)}\right]/\Delta P$, for different compounds are given in Table 6.1. They allow the *Grüneisen parameters* $\Omega_V = (\Gamma_V/\kappa_0) = |x\alpha_V/(\alpha_V - 1)|$ to be obtained for these materials. These are unusually large, and this enhancement can be explained using

Table 6.1 *Grüneisen parameters for different heavy fermions according to the relative pressure variation of several physical quantities shown in Fig. 6.2.* Γ_V *and the compressibility* κ_0 *are in* $Mbar^{-1}$. *The reference pressure for* $CeAl_3$ *is* $P_0 = 1.2$ *kbars otherwise* $P_0 = 0$. *The data for* $CeCu_6$ *is taken from Thompson and Lawrence (1994) and references therein.*

Compound	Γ_V	κ_0	$\Omega_V = \Gamma_V/\kappa_0$	$\alpha_V = V_0/V_c$
$CeAl_3$	89	2.17	41	0.976
$CeRu_2Si_2$	171	0.95	180	0.994
UPt_3	26	0.48	54	0.981
$CeCu_6$	133	1.1	121	0.992

Eq. 6.7 and assuming that α_V is smaller but close to 1 due to the proximity of these systems to a quantum phase transition even at ambient pressure. Assuming that the critical exponent $x = -1$, as will be shown in the next chapter to be the case, the results for Ω_V in Table 6.1 yield values of α_V ranging from $\alpha_V \approx 0.97$ to $\alpha_V \approx 0.99$, as shown in this table. This clearly indicates that the systems we are considering are close to the QCP and a scaling analysis is justified.

The relations between the critical exponents, $2 - \alpha = \nu z$ and $\phi_h = \nu z$, due to the collapse of the data in Fig. 6.2 imply that the free energy obeys a simple one-parameter scaling, $f \propto T_{coh}F[T/T_{coh}, h/T_{coh}]$, where $T_{coh} = |g|^{\nu z}$. This one-parameter scaling is reminiscent of that of the single impurity problem or of other phenomenological, non-critical approaches to the heavy fermion problem. It is surprising that it describes the behaviour of the systems investigated here that are close to a magnetic phase transition. The observation of one-parameter scaling in a three-dimensional critical theory with two independent critical exponents is clearly associated here with the breakdown of the hyperscaling relation $2 - \alpha = \nu(d + z)$. Note that the relation $2 - \alpha = \nu z$ arising from the scaling of the experimental data is just this hyperscaling relation with $d = 0$. It is not surprising, then, that it yields scaling properties which are formally similar to those of a single impurity problem.

Violation of hyperscaling is not uncommon in critical phenomena. It occurs whenever the system is above the upper critical dimension d_c. In this case, for $d > d_c$, as we saw previously, the critical exponents remain fixed at their mean field values and may violate the hyperscaling relation. We have anticipated that this should be a frequent case for quantum phase transitions since the effective dimension, $d_{eff} = d + z$, is increased due to quantum effects and may become larger than d_c. In the next chapter we describe a Gaussian theory of local moments coupled to conduction electrons in the vicinity of a magnetic quantum phase transition. The theory has dynamic exponent $z = 2$ and is in the same universality class as the nearly antiferromagnetic Fermi liquid presented in Chapter 5. In three dimensions both theories are exact as concerns the results for the zero-temperature-critical exponents, since $d_{eff} = d + z > d_c$. They do not lead to one-parameter scaling as observed here, at least very close to the quantum critical point. However, we will show that in these theories there is a region in their phase diagrams where the system is *close but not too close* to the quantum critical point and a one-parameter scaling holds.

6.3 Conclusions

Our analysis of the pressure dependence of several physical quantities for different heavy fermions on a region of the phase diagram, for $(J/W) > (J/W)_c$, $T \leq T_{coh}$ has shown that these systems are close to a quantum critical point and a scaling approach is indeed appropriate. We have obtained, from the experimental data, Grüneisen parameters which are enhanced due to the proximity of these systems to a QCP. We have shown that a one-parameter scaling is sufficient to describe the experiments in the region of the phase diagram probed by the pressure experiments and this is associated with the violation of the hyperscaling relation. This is surprising since, even if these systems are above the upper critical dimension, i.e. $d_{eff} = d + z > d_c$, we would still expect to find at least two independent critical exponents and not a one-parameter scaling. The violation of hyperscaling observed here is different from that we would expect for $d + z > d_c$, where the critical exponents remain fixed at their mean field values. In the present case, the exponents are associated with an Euclidean dimension $d = 0$, suggesting some kind of *local quantum criticality*.

We have shown that, for $z \geq d$, the thermal mass of a Fermi liquid near a quantum instability increases as the QCP is approached. However, this may not be taken as an explanation for the heavy masses of heavy fermions, at least for $3d$ systems, since these materials are nearly antiferromagnetic and should be described by a theory with dynamic exponent $z = 2$. In the next chapter we shall discuss a model for the large masses of heavy fermions consistent with the one-parameter scaling observed in the experiments analysed here.

7

A Microscopic Model for Heavy Fermions

7.1 The Model

In this chapter we will consider a model for heavy fermions introduced by Moriya and Takimoto (1995). It consists of a regular lattice of f-moments which interact with an electron gas and through this gas with themselves. The Hamiltonian which describes the interaction between the f-moments is given by

$$H = \sum_{ij} J_{ij} \vec{S}_i . \vec{S}_j = \sum_q J_q |\vec{S}_q|^2, \tag{7.1}$$

where the coupling J_{ij} arises due to the interaction J between the local moments and the conduction electrons as in the Kondo lattice Hamiltonian, Eq. 6.2. The interaction of the f-moments with the electron gas is fully described by a local dynamic susceptibility,

$$\chi_L(\omega) = \frac{\chi_L}{1 - i\omega/\Gamma_L},$$

where Γ_L and χ_L are local parameters related to the Kondo temperature T_K of a single f-moment imbedded in the electron gas. In fact, $\Gamma_L \propto T_K$, but the product $\Gamma_L \chi_L = constant$, independent of T_K (Moriya and Takimoto, 1995).

The free energy of this model in the Gaussian approximation is given by

$$f_{sf} = -\frac{3}{\pi} \sum_{\vec{q}} T \int_0^\infty \frac{d\lambda}{e^\lambda - 1} \tan^{-1}\left(\frac{\lambda T}{\Gamma_q}\right), \tag{7.2}$$

where

$$\Gamma_q = \Gamma_L(1 - J_Q \chi_L) + \Gamma_L \chi_L A q^2$$

and J_Q is the q-dependent exchange coupling between the local f-moments at the wavevector Q of the incipient magnetic instability. A is the *stiffness* of the lifetime of the spin fluctuations defined by the small wave vector expansion of the magnetic coupling close to the wave vector Q, i.e.

$$J_Q - J_{Q+q} = Aq^2 + \cdots . \tag{7.3}$$

Then, Γ_q can be rewritten as

$$\Gamma_q = \Gamma_L \chi_L A \xi^{-2}[1 + q^2 \xi^2],$$

where the correlation length, $\xi = (A/|J_Q - J_Q^c|)^{1/2} = \sqrt{A/|g|}$ diverges at the critical value of the coupling, $J_Q^c = \chi_L^{-1}$, with the Gaussian critical exponent, $\nu = 1/2$. Then the present microscopic model describes a quantum phase transition from a non-magnetic to a magnetic ground state characterised by the wavevector Q, the quantum critical point being located at $J_Q = J_Q^c$. The free energy can be rewritten as

$$f_{sf} = -\frac{3}{\pi} \sum_{\vec{q}} T \int_0^\infty \frac{d\lambda}{e^\lambda - 1} \tan^{-1} \left(\frac{\lambda T \xi^z}{\Gamma_L \chi_L A(1 + q^2 \xi^2)} \right), \tag{7.4}$$

where we have identified the value of the dynamic critical exponent of the theory, $z = 2$, to write the free energy in terms of the scale-invariant form of temperature, $T\xi^z$. Replacing the sum in $\vec{q}$ by an integral in $\int_0^{q_c} d\vec{q}$ and introducing the variables $x = q\xi$, we can write the free energy in the scaling form

$$f_{sf} = |g|^{\nu(d+z)} F[T/T_{coh}]$$

in the limit $q_c \xi \to \infty$. The *coherence temperature*, T_{coh} (Continentino, 2000), is given by

$$k_B T_{coh} = A\Gamma_L \chi_L \xi^{-z} = \Gamma_L \chi_L |g|^{\nu z} = \Gamma_L \chi_L |J_Q - J_Q^c|$$

and is *independent* of the stiffness A. Furthermore, since it depends on the product $\Gamma_L \chi_L$, *it is a new energy scale independent of the Kondo temperature T_K*. The exponent $\nu z = 1$ ($\nu = 1/2$ and $z = 2$). This model is in the same universality class as the Gaussian action for a nearly antiferromagnetic Fermi liquid discussed in Section 5.5.

The argument of the $\tan^{-1}$ in Eq. 7.4 can be written as

$$\left(\frac{\lambda T \xi^z}{A\Gamma_L \chi_L} [1 - \frac{q^2 \xi^2}{1 + q^2 \xi^2}] \right).$$

The term in brackets $[\cdots]$ is always ≤ 1, furthermore the exponential in Eq. 7.4 cuts off the contribution for the integral from large values of λ, consequently for

$(T\xi^z/\Gamma_L\chi_L A) = (T/T_{coh}) << 1$, we can expand the $\tan^{-1}$ for small values of its arguments. The specific heat, $C/T = -\frac{\partial^2 f_{sf}}{\partial T^2}$, for $T << T_{coh}$, is given by

$$C/T = \frac{\partial^2}{\partial T^2}\left[\frac{3T^2}{\pi T_{coh}}\frac{V}{(2\pi)^3}\int_0^\infty \frac{d\lambda\lambda}{e^\lambda - 1}\int_0^{q_c}\frac{d\vec{q}}{1+q^2\xi^2}\right] \qquad (7.5)$$

Since the right hand side of this equation is independent of temperature, we have that, for $T << T_{coh}$, the specific heat is linear in temperature as in a Fermi liquid. This is in agreement with our previous interpretation of the coherence temperature using the scaling approach. It is the characteristic temperature, below which, the free energy is quadratic in temperature and consequently the system exhibits Fermi liquid behaviour.

We can specialise Eq. 7.5 for $d = 3$ and rewrite it as

$$C/T = \frac{\partial^2}{\partial T^2}\left[\frac{\pi T^2\xi^{(z-d)}}{2\Gamma_L\chi_L A}\frac{4\pi V}{(2\pi)^3}\int_0^{q_c\xi}\frac{dy y^2}{1+y^2}\right], \qquad (7.6)$$

which yields

$$C/T = \frac{\pi\xi^{(z-d)}}{\Gamma_L\chi_L A}\frac{4\pi V}{(2\pi)^3}q_c\xi\left(1 - \frac{\tan^{-1}q_c\xi}{q_c\xi}\right). \qquad (7.7)$$

In the limit $q_c\xi >> 1$ using $\Gamma_L\chi_L = 1/2\pi$, we obtain the result of Moriya and Takimoto (1995):

$$\frac{C}{T} = \frac{6\pi^2 N k_B^2}{A q_c^2}, \qquad (7.8)$$

a non-universal, cut-off-dependent value for the thermal mass m_T. In the opposite limit, i.e. $q_c\xi << 1$, since $\tan^{-1}y \approx y - y^3/3 + y^5/5 + \cdots$ for small y, we get for the specific heat

$$C/T = \frac{\pi\xi^{(z-d)}}{\Gamma_L\chi_L A}\frac{4\pi V}{(2\pi)^3}q_c\xi\left[\frac{1}{3}(q_c\xi)^2 - \frac{1}{5}(q_c\xi)^4 + \cdots\right]. \qquad (7.9)$$

The first term is independent of A and yields

$$C/T = \frac{\pi N k_B^2}{\Gamma_L\chi_L}\frac{1}{|J_Q - J_Q^c|} = \frac{\pi N k_B}{T_{coh}}. \qquad (7.10)$$

7.2 Local Quantum Criticality

We could get the result above for the specific heat directly using a *local free energy* obtained from Eq. 7.4, neglecting the q-dependence of Γ_q and taking $\sum_{\vec{q}} \to N$. We find

$$f_{sf}^L = -\frac{3}{\pi}NT\int_0^\infty \frac{d\lambda}{e^\lambda - 1}\tan^{-1}\left(\frac{\lambda T}{T_{coh}}\right). \qquad (7.11)$$

Table 7.1 *Parameters for some heavy fermion systems. All experimental data are taken from Thompson and Lawrence (1994) and references therein. (∗) obtained from Eq. 7.10 for the specific heat. (∗∗) along the c-axis. (∗ ∗ ∗) in the basal plane. For the transport parameter, the average number of conduction electrons per atom, n = 1.*

	$CeRu_2Si_2$	$CeCu_6$	UPt_3
$\mu(\mu_B)$	2.5	2.5	3.0
$v_0^{1/3}(\times 10^{-8}cm)$	4.4	4.7	4.1
$T_{coh}(K)^*$	67	15	58
$\chi/\mu^2\ (\times 10^{35}erg^{-1}cm^{-3})$	5.8	7.9	2.5
$\frac{C/T}{\pi^2 k_B^2}\ (\times 10^{35}erg^{-1}cm^{-3})$	3.9	13.0	5.6
$\frac{\chi/\mu^2}{C/T\pi^2 k_B^2}$	1.46	0.59	0.44
$A_R\ (\mu\Omega cm K^{-2})$	0.40	14.4	0.7** - 1.6***
$\frac{A_R}{(C/T)^2}\ (\times 10^{-5}\mu\Omega cm(\frac{molK}{mJ})^2)$	0.27	0.51	0.34** - 0.79***
$(\Gamma_L\chi_L)(J/W)^2$	0.40	0.71	0.54** - 1.25***
(J/W)	0.47	0.63	0.39** - 0.59***

This free energy can be written in the scaling form $f_{sf}^L = |g|^{\nu z} F[T/T_{coh}]$, such that $2 - \alpha = \nu z$. If we use the hyperscaling relation, $2 - \alpha = \nu(d + z)$, it is satisfied for an Euclidean dimension $d = 0$, as expected for a local, q-independent theory. It may also be written as

$$f_{sf}^L = T G[T/T_{coh}] \qquad (7.12)$$

in a way that emphasises its one-parameter scaling form. This is a local critical theory since the characteristic temperature T_{coh} is a direct consequence of the quantum phase transition.

Let us examine the condition $q_c\xi \ll 1$ that limits the regime of local quantum criticality. This may be written as, $q_c\sqrt{A/|g|} \ll 1$, which can be satisfied *either because the system is distant from criticality, i.e. |g| is large, or because A is small.* Rewriting this condition as $q_c/\sqrt{|g|} \ll 1/\sqrt{A}$, we notice that, when $A \to 0$, it holds *arbitrarily close* to the QCP, i.e. for g arbitrarily small. For heavy fermions with large masses this is expected to be the case, as we show below.

In Table 7.1 we give values for the coherence temperature, obtained from Eq. 7.10 and the measured linear term of the specific heat, for some heavy fermion systems.

Susceptibility and Wilson Ratio

The zero temperature uniform susceptibility of the system near the antiferromagnetic instability, in the limit $q_c\xi << 1$, can be directly obtained from the magnetic field (h) dependent, $T = 0$, q-independent free energy:

$$f_{sf}^L = -\frac{3N}{2\pi} \int_0^{\omega_c} d\omega \tan^{-1} \left[\frac{\omega + \mu h}{\Gamma_L \chi_L |J_Q - J_Q^c|} \right]. \tag{7.13}$$

Notice that the magnetic field in this expression scales as (h/h_c) where the characteristic field, $h_c \propto T_{coh}$. Integrating, differentiating twice, taking the value at $h = 0$ and the limit $\omega_c \to \infty$, we obtain

$$\chi_0 = -\left(\frac{\partial^2 f_{sf}^L}{\partial h^2} \right)_{h=0} = \frac{3N\mu^2}{2\pi \Gamma_L \chi_L} \frac{1}{|J_Q - J_Q^c|} = \frac{3N\mu^2}{2\pi k_B T_{coh}}. \tag{7.14}$$

From the general expression for the uniform susceptibility, $\chi^{-1} = \chi_Q^{-1} + Aq_c^2$, where χ_Q is the staggered susceptibility, we obtain that in the *local limit*, $q_c\xi << 1$, $\chi = \chi_Q = \chi_0$. Then the staggered and uniform susceptibilities coincide in this limit, being equally enhanced. The Wilson ratio in the same limit is given by (Continentino, 2000)

$$\frac{\chi_0/\mu^2}{C/\pi^2 k_B^2 T} = \frac{3}{2} = 1.5, \tag{7.15}$$

which turns out to be a universal number since the dependence on the distance to the critical point, $|J_Q - J_Q^c|$ and on the dimensionless quantity $\Gamma_L \chi_L$ cancels out. We emphasise that the above results are valid in the regime, $q_c\xi << 1$; that is, if the system satisfies the condition, $q_c/\sqrt{|J_Q - J_Q^c|} << 1/\sqrt{A}$. The experimental values given in Table 7.1 are in fair agreement with the result of the theory in this limit, specially those for $CeRu_2Si_2$.

Resistivity and Kadowaki–Woods Ratio

The resistivity due to spin fluctuations in the regime $q_c\xi << 1$ can be calculated using standard methods and is given by (Lederer and Mills, 1968)

$$\rho = \rho_0 \frac{1}{T} \int_0^\infty d\omega \frac{\omega \Im m \chi_Q(\omega)}{(e^{\beta\omega} - 1)(1 - e^{-\beta\omega})}, \tag{7.16}$$

where

$$\Im m \chi_Q(\omega) = \chi_Q^s \frac{\omega \xi_L^z}{1 + (\omega \xi_L^z)^2}$$

with

$$\chi_Q^s = \frac{1}{|J_Q - J_Q^c|}$$

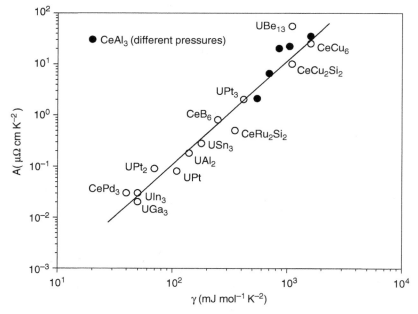

Figure 7.1 The coefficient A of the T^2 term of the resistivity versus the coefficient γ of the linear term of the specific heat for several heavy fermions (Continentino, 1994 and references therein). The straight line is the equation $A = b\gamma^2$ with $b = 1.1 \times 10^{-5}$ $(\mu\Omega cm K^2)/(mJ^2 mol^{-2})$.

and

$$\xi_L^z = \frac{\chi_Q^s}{\Gamma_L \chi_L}.$$

The quantity ρ_0 is given by

$$\rho_0 = \left(\frac{J}{W}\right)^2 \frac{m_c}{N_c e^2 \tau_{Fc}} (N/N_c),$$

where J is the coupling constant per unit cell between localised and conduction electrons. W, m_c and N_c are the bandwidth, the mass and the number of conduction electrons per unit volume with Fermi momentum k_{Fc}, such that, $\hbar \tau_{Fc}^{-1} = \hbar^2 k_{Fc}^2/2m_c$ and N is the number of atoms per unit volume.

Using the definitions above, we can rewrite the spin-fluctuation resistivity as $\rho = \rho_0 \Gamma_L \chi_L R(\tilde{T})$, where

$$R(\tilde{T}) = \frac{1}{\tilde{T}} \int_0^\infty d\tilde{\omega} \frac{1}{(e^{\tilde{\omega}/\tilde{T}} - 1)(1 - e^{-\tilde{\omega}/\tilde{T}})} \frac{\tilde{\omega}^2}{1 + \tilde{\omega}^2} \qquad (7.17)$$

with $\tilde{\omega} = \omega \xi_L^z$ and $\tilde{T} = T \xi_L^z$. For, $T \ll T_{coh} = (\Gamma_L \chi_L / k_B)|J_Q - J_Q^c|$, we have $R(T \ll T_{coh}) \approx \frac{\pi^2}{3}(\frac{T}{T_{coh}})^2$ and finally

$$\rho(T << T_{coh}) = \rho_0 \Gamma_L \chi_L \frac{\pi^2}{3} (\frac{T}{T_{coh}})^2 = A_R T^2 \qquad (7.18)$$

where

$$A_R = \frac{\rho_0 \pi^2}{3} \frac{k_B^2}{\Gamma_L \chi_L} \frac{1}{|J_Q - J_Q^c|^2}. \qquad (7.19)$$

Notice that the same coherence temperature, as defined through the thermodynamic quantities, appears in the transport properties. In this case T_{coh} is the characteristic temperature below which the resistivity varies quadratically with temperature, as appropriate to a FL, and that sets the scale for this contribution. For $T >> T_{coh}$, $R(T) \approx \frac{\pi T}{2 T_{coh}}$ and the resistivity varies linearly with temperature. It is given by

$$\rho(T >> T_{coh}) = \rho_0 \Gamma_L \chi_L \frac{\pi}{2} \frac{T}{T_{coh}}.$$

The Kadowaki–Woods ratio (Kadowaki and Woods, 1986), $A_R/(C/T)^2$, is given by

$$\frac{A_R}{(C/T)^2} = \frac{\rho_0 \Gamma_L \chi_L}{3(N k_B)^2}, \qquad (7.20)$$

which depends on the local parameters, $\Gamma_L \chi_L$, and consequently on the nature of the magnetic ion ($4f$, $5f$ or d, for example), but *not* on the distance to the critical point, $|J_Q - J_Q^c|$. From the equation above and the expression for ρ_0, we obtain $(\Gamma_L \chi_L) [\frac{J}{nW}]^2 = \frac{6.55 \times 10^{-3}}{n^{2/3} v_0^{1/3}} \frac{A_R}{(C/T)^2}$, with $n = (N_c/N)$, the average number of conduction electrons per atom and $v_0^{1/3}$, the average atomic radius given in cm. The values of $(\Gamma_L \chi_L)(J/W)^2$ obtained from the experimental results and with $n = 1$, are given in Table 7.1. If we use $(\Gamma_L \chi_L) = 1/2\pi$, the value for $S = 1/2$ (Takimoto and Moriya, 1996), we get $(J/W) = 1.6$, 1.8 and 2.1 for $CeRu_2Si_2$, UPt_3 and $CeCu_6$, respectively, which are too large. On the other hand, extending the Korringa relation (Takimoto and Moriya, 1996) for arbitrary spin, i.e. $\Gamma_L \chi_L = 2(\mu/g_J \mu_B)^2/3\pi$, where μ is the experimental magnetic moment (in μ_B) given in Table 7.1, we get the values for (J/W) also shown in this Table ($n = 1$). These values are in agreement with those expected for non-magnetic heavy fermions (Coqblin, 1977), although the value of $(J/W)_c$ separating the magnetic from the Fermi liquid ground states in heavy fermions is still unknown. Takimoto and Moriya (1996) have also calculated the Kadowaki–Woods ratio for heavy fermions and obtained that, not too close to the QCP, it is nearly independent of $|g|$.

Local Regime and One-Parameter Scaling

The theory presented above, in the local regime, i.e. for $q_c \xi \ll 1$, gives the correct description of the one-parameter scaling observed in the pressure experiments presented and analysed in Chapter 6 (Continentino, 1998). We have shown that in this regime, $C/T \propto T_{coh}^{-1}$, $\chi_0 \propto T_{coh}^{-1}$, $A_R \propto T_{coh}^{-2}$ and $h_c \propto T_{coh}$, such that, a single characteristic temperature, T_{coh}, controls the scaling of all these quantities as found experimentally in the pressure experiments. These experiments, together with the theoretical results of this section, allow the crossover exponent $\nu z = 1$ to be obtained, which was assumed for the calculation of the enhancement factors of the Grüneisen parameters in the previous chapter. The regime of local quantum criticality may be summarised by the scaling form of the free energy:

$$f_L \propto T_{coh} F[T/|T_{coh}, h/T_{coh}],$$

with $T_{coh} = |g|^{\nu z}$ and $\nu z = 1$. If we define an *effective* exponent α_L in this regime by $f_L \propto |g|^{2-\alpha_L}$, it takes the value $\alpha_L = 1$ and the experimental relation, $2 - \alpha = \nu z$, Eq. 6.8 of Chapter 6 is immediately satisfied with $\alpha = \alpha_L$. Also, Eq. 6.9 of the previous chapter is in accordance with the result $h_c \propto T_{coh}$ obtained in this section. Besides, in the local regime, T_{coh} as obtained from the T^2 term of the resistivity coincides with the thermodynamic coherence temperature as observed. Then, together, theory and experiment allow us to conclude that the heavy fermions, $CeRu_2Si_2$, $CeCu_6$, UPt_3 and $CeAl_3$ under the physical conditions of Chapter 6 are in a regime of local quantum criticality. They are close, *but not too close*, to the QCP, such that the condition $q_c\xi \ll 1$ or $q_c/\sqrt{|g|} << 1/\sqrt{A}$ is satisfied (see Fig. 7.2). Notice that for the compounds used in the experiments, pressure drives them further away from the critical point, i.e. deeper in the local regime $q_c\xi > 1$.

It is interesting that the exponents of the theory, $\nu z = 1$ satisfy the hyperscaling relation $2 - \alpha_L = \nu(d+z)$ in the local regime for $d = 0$. This is consistent with the

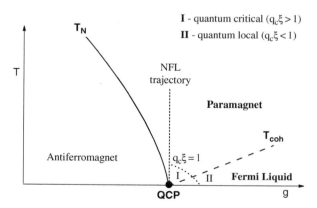

Figure 7.2 The phase diagram of a nearly antiferromagnetic Fermi liquid in $3d$. The dotted line $q_c\xi = 1$ separates the true critical regime $q_c\xi >> 1$ from the local regime $q_c\xi \sim 1$.

nature of this regime. In spite of the spatial correlations being short-ranged in the local regime, time fluctuations are critically correlated and yield a one-parameter scaling, as we have shown. This one-parameter scaling is a direct consequence of the quantum character of the transition. Formally, the local regime is equivalent to a problem in Euclidean dimension $d = 0$, but with effective dimensionality $d_{eff} = z = 2$ associated with the *time directions*. The crossover from the local regime to the true critical regime at $q_c \xi = 1$ is smooth and can be seen as a dimensional crossover from $d = 0$ to $d = 3$ (see Fig. 7.2).

7.3 Critical Regime

As the quantum critical point is approached and $q_c \xi \to \infty$, the full q-dependence of the dynamic susceptibility must be taken into account. In this critical region of the phase diagram, the thermal mass and the uniform susceptibility of the Fermi liquid below the coherence line, at $d = 3$, are given by

$$C/T = 6\pi^2 N k_B^2 / A q_c^2 + O(|g|^{1/2})$$
$$\chi_0 = N\mu^2 / A q_c^2 \tag{7.21}$$

and both saturate at non-universal, i.e. cut-off-dependent, values and do not scale in terms of the distance to the QCP, although of course A may depend on pressure. The Wilson ratio $WR = 1/6$ (see Eq. 7.15:) much smaller than in the local regime. The scaling contribution for the thermal mass of a nearly antiferromagnetic system in the Fermi liquid regime below the coherence line is given by

$$m_T = C/T \propto |g|^{\nu(d-z)} \propto \xi^{(z-d)}.$$

For $z = 2$, in $3d$, with the Gaussian exponent $\nu = 1/2$ this yields $m_T \propto \sqrt{|g|}$. In the present theory, this contribution comes from the second term, proportional to $\tan^{-1}$, in Eq. 7.7. Since it vanishes at the QCP ($g = 0$), the thermal mass close to the AFQCP is determined by a non-universal contribution, in the present theory by Eqs. 7.8 or 7.21, which correspond to the first term in Eq. 7.7. Then, in this approach, the large masses of heavy fermions arise from the small values for the stiffness A.

In the critical regime, i.e. for $q_c \xi \gg 1$ and below the coherence line, the resistivity also has a quadratic temperature dependence, $\rho = A_R^M T^2$, at low temperatures, but the coefficient $A_R^M \propto \chi_Q^{1/2} \propto |J_Q - J_Q^c|^{-1/2}$ where χ_Q is the staggered susceptibility which diverges with the Gaussian exponent $\gamma = 1$ close to the QCP. Consequently, in this critical regime, although the resistivity varies quadratically with temperature, the coefficient of this term does not scale as T_{coh}^{-2}, in disagreement with experiments in the heavy fermion systems studied in Chapter 6.

It is clear from the results above that the present Gaussian theory in the critical, q-dependent regime, does not yield the one-parameter scaling obtained in the pressure experiments that we analysed in the previous chapter, even though it yields the correct description of the quantum critical behaviour for $d + z > 4$. In this critical regime, C/T and χ_0, as shown in Eqs. 7.21, are controlled by the *stiffness A*. The large thermal masses of heavy fermions in this model are due to the smallness of the stiffness A which in turn implies the flatness of the spin-fluctuation spectrum near the instability wave vector Q (Lacroix and Pinettes, 1995). On the other hand, in the local limit, $q_c \xi \ll 1$ of the theory, the relevant physical properties in the Fermi liquid regime are universal, in the sense that they are independent of the cut-off q_c and the stiffness A. In this case they are exclusively determined by the coherence temperature (Eqs. 7.10, 7.14 and 7.19) (Continentino, 2000).

The theory presented in this chapter is in the same universality class as the theory of the *nearly antiferromagnetic Fermi liquid*, presented in Chapter 5. They are both Gaussian theories with dynamic exponent $z = 2$ and consequently yield identical results in the scaling regime. The differences arise from the microscopic origin of the parameters of these models. Besides, it is clear from the results of previous chapters that these theories yield the correct description of the quantum critical behaviour of systems close to the antiferromagnetic instability whenever $d + z \geq 4$.

A further result of the theories of Moriya and Takimoto (1995) and the nearly AF Fermi liquid is that they predict the existence of a critical line of finite temperature phase transitions, for $d > 2$, in the magnetic side of the phase diagram. Such a line close to the quantum critical point is described by $T_N \propto |g|^\psi$ where the shift exponent, in both cases is given by $\psi = (d + z - 2)/z$ with $z = 2$. We emphasise that although these Gaussian theories describe correctly the quantum critical behaviour for $d + z \geq 4$, this is not the case for the finite temperature transitions at T_N. In $d = 3$, these transitions are governed by *Wilson exponents* and not Gaussian ones.

7.4 Generalised Scaling and the Non-Fermi Liquid Regime

An interesting and deep consequence of the scaling approach and the interpretation of the coherence temperature as given by Eq. 6.4 is that, if the system is at the quantum critical point, i.e. $|g| = 0$, on reducing the temperature it never crosses the coherence line and consequently never enters the Fermi liquid regime. Then, $|g| = 0$ is a *non-Fermi liquid* (NFL) trajectory in the phase diagram of Fig. 7.2 (Continentino, 1991). For a Gaussian model with $d + z > d_c$, the thermodynamic properties along this non-Fermi liquid trajectory can be easily obtained using naive scaling. The correlation length at the QCP, $g = 0$, increases with decreasing temperature as

$$\xi \propto T^{-1/z}. \tag{7.22}$$

The staggered susceptibility along the QCT diverges as

$$\chi_s \propto T^{-\gamma/\nu z}, \tag{7.23}$$

where the exponents take Gaussian or mean field values. The specific heat along the NFL trajectory is given by

$$C/T \propto T^{\frac{d-z}{z}}.$$

For the model of the previous sections or for the nearly antiferromagnetic metal with $z = 2$ in $3d$ these equations yield $\xi \propto T^{-1/2}$, $\chi_s \propto T^{-1}$ and $C/T = \gamma_0 - a\sqrt{T}$, where γ_0 is a non-universal contribution obtained previously.

We have seen in previous chapters that for $d + z > 4$, although Gaussian theories, as that presented in this chapter, describe correctly the critical exponents associated with the quantum phase transition, they do not yield the correct temperature behaviour, especially at quantum criticality ($g = 0$). The reason is that the quartic interaction u, in spite of being irrelevant at the QCP as concerns the values of the critical exponents, is *dangerously irrelevant* and can modify the critical behaviour especially along the QCT. As a first consequence of this character of u, the shift exponent that determines the shape of the critical line of finite temperature phase transitions, $T_N = (1/u)|g|^\psi$, near the QCP, is modified with respect to that expected from the generalised scaling hypothesis $\psi = \nu z$ and is given by

$$\psi^{-1} = (d + z - 2)/z. \tag{7.24}$$

Also, the correlation length ξ and the staggered susceptibility χ_s have their temperature dependence along the QCT altered by the dangerously irrelevant quartic interaction

$$\xi = \frac{1}{\sqrt{u}} T^{\frac{-\nu}{\psi}} \tag{7.25}$$

and

$$\chi_s = \frac{1}{u} T^{\frac{-\gamma}{\psi}}, \tag{7.26}$$

with ψ given by Eq. 7.24. This new type of behaviour arises when u is taken into account, either perturbatively, as in the theories of Moriya and Takimoto (1995), or using the renormalisation group as for the nearly antiferromagnetic Fermi liquid of Millis (1993) presented in Chapter 5. Alternatively, these results can be easily obtained using the generalised scaling theory presented in Section 1.7 of Chapter 1, assuming that the thermal and quantum exponents take Gaussian values, i.e. $\nu = \tilde{\nu} = 1/2$ and $\gamma = \tilde{\gamma} = 1$. The same assumption leads to the scaling contribution for the specific heat along the NFL

$$C/T \propto u^{3/2} T^{5/4},$$

for a nearly antiferromagnetic system with Gaussian thermal and quantum exponents. We took $d = 3$, $\nu = 1/2$ and $z = 2$ such that $\psi = 2/3$.

It is easy to check that the expressions for the correlation length and staggered susceptibility modified by the quartic interaction are more singular than those obtained from the purely Gaussian approach with $u = 0$. They should dominate the thermal behaviour along the NFL trajectory. This is not the case for the specific heat for which the Gaussian contribution is more singular.

Another point we can analyse here is related to the so-called ω/T scaling. As we saw in Chapter 1, frequency near a quantum phase transition enters the scaling functions in the scaling form, $\omega\xi^z$. For a purely Gaussian model along the quantum critical trajectory, one can immediately see from Eq. 7.22 for the correlation length that it leads to ω/T scaling. However, in the presence of a quartic interaction u for $d + z \geq d_c$, this scaling form is modified and one gets $\omega/T^{-(\nu z/\psi)}$, as can be checked from Eq. 7.25 for the correlation length. Notice that violation of ω/T scaling is closely related to the breakdown of the generalised scaling hypothesis $\psi = \nu z$ by the dangerously irrelevant interaction u.

7.5 Local Regime near the QCP

The local limit is not useful at quantum criticality, $g = 0$, since its characteristic energy given by T_{coh} vanishes and it gives only trivial results. However, a curve $q_c\xi = 1$ can still be defined all over the non-critical region of the phase diagram, marking the onset of truly quantum critical behaviour. For example, along the QCT there is a characteristic temperature T_L, defined by, $q_c\xi(T_L) = 1$, which separates the local from the critical regime. Only for $T \ll T_L$ is i the true critical regime attained. For the Gaussian, $z = 2$, quantum critical point, using Eq. 7.25 for the correlation length, we obtain the following condition to enter this critical regime, $q_c\xi \gg 1$:

$$\frac{q_c}{\sqrt{u}} T^{-3/4} \gg 1,$$

where, for the nearly antiferromagnetic metal model, u is the dangerously irrelevant quartic interaction. Even without an estimate for the quantity u in heavy fermions, just from the fact that the power law exponent governing the temperature dependence of the correlation length is smaller than 1, we may expect that the condition above is satisfied only at very low temperatures. In fact, neutron scattering experiments suggest that a local scaling description of the magnetic behaviour of at least one heavy fermion at the quantum critical point holds, down to very low temperatures (Schröder, Aeppli, Bucher *et al.*, 1998).

7.6 Quantum Lifshitz Point

We have obtained that the stiffness A in Eq. 7.3 must be small in heavy fermions and this is connected to the large thermal masses of these materials. We may wonder: what is the influence of, for example, the quartic term in the expansion of the q-dependent interaction

$$J_Q - J_{Q+q} = Aq^2 + Bq^4. \tag{7.27}$$

In this case, the relevant modification in the calculation of the free energy, Eq. 7.4, appears in the argument of the $\tan^{-1}$, namely

$$y = \frac{(\lambda T \xi^z / \Gamma_L \chi_L A)}{1 + q^2\xi^2 + (Bq^2/A)q^2\xi^2}$$

with ξ as defined before and $z = 2$. Since the maximum value of q is q_c, we certainly overestimate the value of the B term, compared to the A term, if we write

$$y = \frac{\lambda(T/T_{coh})}{1 + q^2\xi^2 + (Bq_c^2/A)q^2\xi^2}$$

with T_{coh} as defined previously. We rewrite this equation as

$$y = \frac{\lambda(T/T_{coh})}{1 + [1 + (Bq_c^2/A)]q^2\xi^2},$$

which leads us to define a new correlation length

$$\xi' = \xi\sqrt{1 + \frac{Bq_c^2}{A}}$$

such that the condition $q_c\xi' < 1$, may be written as

$$\sqrt{\frac{Aq_c^2 + Bq_c^4}{|g|}} < 1$$

or still, taking into account the other terms of the expansion for $J_{Q+q_c} - J_Q$,

$$\sqrt{\frac{J_{Q+q_c} - J_Q}{|g|}} < 1.$$

This means that the local model holds whenever the distance to the QCP becomes larger, in energy units, than the total spread of the spectrum of spin fluctuations.

As long as the stiffness $A > 0$, the effect of the other terms in the expansion, Eq. 7.27, is just to modify the criterion for the validity of the local model, i.e. the condition $q_c\xi \ll 1$, from the previous form, $Aq_c^2 \ll |g|$, to $(J_{Q+q_c} - J_Q) \ll |g|$. This condition implies that the local regime holds whenever the system is sufficiently far away from the QCP, such that $|g|$ is larger than the spreading of the spectrum of spin fluctuations up to the cut-off q_c.

The situation is completely different if the coefficient $A = 0$ (Ramazashivilii, 1999). In this case, the universality class of the antiferromagnetic quantum critical point is altered. Let us consider first an anisotropic situation where the magnetic correlations along planes are different from those along the *soft* perpendicular direction. In this case the Gaussian free energy can be written as

$$ f \propto T \int_0^\infty \frac{d\lambda}{e^\lambda - 1} \int d^{d-1}q_\parallel \int dq_\perp \tan^{-1}\left(\frac{\lambda(T/T_{coh})}{(1 + q_\parallel^2 \xi_\parallel^2 + q_\perp^4 \xi_\perp^4)} \right) $$

with $T_{coh} \propto |g|^{\nu_\parallel z}$, where $\nu_\parallel z = 1$. Also $\xi_\parallel \propto |g|^{-1/2}$, such that $\nu_\parallel = 1/2$ and $\xi_\perp \propto |g|^{-1/4}$, which implies $\nu_\perp = 1/4$. The scaling form of the free energy in this case is given by

$$ f \propto |g|^{\nu_\parallel z + \nu_\parallel(d-1) + \nu_\perp} F\left[\frac{T}{T_{coh}}\right]. $$

In the Fermi liquid regime $T \ll T_{coh}$, this approach leads to a scaling contribution to the thermal mass vanishing as $m_T = C/T \propto |g|^{1/4}$. At the critical point, $|g| = 0$, the scaling contribution to the free energy is obtained assuming that the scaling function in the equation above, $F[t] \propto t^x$, for $t \to \infty$, and requiring, as usual, that the dependence of the free energy on $|g|$ cancels out. This yields $f \propto T^{9/4}$ and $C/T \propto \partial^2 f/\partial T^2 \propto T^{1/4}$. The expression for the critical line is obtained at the point where the coefficient of the $q_\parallel^2$ term vanishes. In this case it is easy to check that $z = 4$, such that, $\nu_\perp z = 1$. The shift exponent is then given by $\psi^{-1} = z/(d + z - 2) = 4/(3 + 4 - 2) = 4/5$.

Let us return to the isotropic situation and assume that the first non-null coefficient of the expansion of Eq. 7.27 in powers of q is A_n such that $J_{Q+q_c} - J_Q = A_n q^n + A_{n+2} q^{n+2} + \cdots$. It is straightforward to verify in this case that the correlation length exponent $\nu = 1/n$ but $\nu z = 1$ such that the dynamic exponent $z = n$. The thermal mass is enhanced according to $C/T \propto |g|^{(\frac{d}{n} - 1)}$ for $T \ll T_{coh}$ and at the QCP, $C/T \propto T^{\frac{d-n}{n}}$.

7.7 Conclusions

In this chapter, we have presented a microscopic theory of heavy fermions which describes a quantum phase transition from a paramagnetic state at $T = 0$ to a spin-density wave ordered state characterised by a wave vector Q. The theory is Gaussian, with a dynamic exponent $z = 2$, and close to the zero temperature phase transition it yields results which are similar to those of the nearly antiferromagnetic Fermi liquid since both theories are in the same universality class. As a consequence of the results of previous chapters, it is clear that the present theory yields the correct description of the quantum critical behaviour near the Q-wave instability for $d + 2 \geq 4$, which is the case of two and three dimensions. The theory yields results along the non-Fermi liquid trajectory and for the shift exponent which are in good agreement with experiments in heavy fermions, in particular, if we accept that the critical behaviour of some of these systems is governed by two dimensional fluctuations. In this case it yields the ubiquitous $\ln T$ dependence of C/T along the trajectory $|g| = 0$.

Away from the QCP, in the non-critical side of the phase diagram, the theory does not describe a one-parameter scaling, at least in the critical regime $q_c \xi \gg 1$. We have shown, however, that not too close to the QCP, for $q_c \xi \leq 1$, the present approach is equivalent to a local model where fluctuations, although not correlated along the spatial dimensions, are critically correlated along the time directions due to the quantum character of the phase transition. The local regime of the theory yields the one-parameter scaling observed in heavy fermions. The relevant thermodynamic and transport properties in this local regime are determined by the coherence temperature. We can conclude that the heavy fermions studied in Chapter 6 and which obey one-parameter scaling are close, *but not too close* to the quantum critical point such that they are described by the local regime of the model presented in this chapter.

8

Metal and Superfluid–Insulator Transitions

8.1 Conductivity and Charge Stiffness

In this chapter, we will consider initially metal–insulator (MI) transitions which occur in pure systems without disorder. In the final section we consider a superfluid–insulator transition induced by some special type of disorder. We will show that the quantities which characterise the phase transitions in both problems obey similar scaling laws and are governed by similar exponents.

Differently from the magnetic transitions treated before metal–insulator (MI) transitions have no obvious order parameter to distinguish the metallic from the insulating phase. This precludes a power expansion of the free energy in terms of a small quantity as for magnetic phase transitions. In spite of that, the concepts of phase transitions and critical phenomena turn out to be very useful to describe metal–insulator transitions. In case this problem is approached using the renormalisation group, the localisation transition is associated with an unstable fixed point and the flow of the *RG* equations to the different attractors is sufficient to distinguish the nature of the phases. Then, in general, we can associate a set of critical exponents with the MI transitions. Besides, these exponents are not independent but obey scaling relations. Since we are concerned with zero temperature instabilities, the quantum hyperscaling law, Eq. 1.18, plays a crucial role in this problem.

We will distinguish here two different types of metal–insulator transitions. Those due to a competition between parameters of the relevant Hamiltonian and metal–insulator transitions arising by varying the number of electrons or chemical potential. An example of the former is the Mott transition due to the competition between kinetic energy and the local Coulomb repulsion as described by the half-filled band Hubbard model. The latter will be referred as *density-driven metal–insulator transitions* in analogy with the superfluid–insulator transition in bosonic systems with varying particle number (Fisher *et al.*, 1989). These two types of transitions may occur in the same system or Hamiltonian. They in general belong to different universality classes and consequently have different critical exponents.

We start discussing how to characterise a metal–insulator transition and define the relevant critical exponents. The physical quantity which can be unambiguously related to the localisation phenomenon is the *conductivity effective mass* m_c, obtained from the frequency dependent conductivity $\sigma(\omega)$ or alternatively, the *charge stiffness* $D_c \propto 1/m_c$. For a perfect conductor, as a lattice translation invariant system at $T = 0$, the frequency dependent conductivity $\sigma^{xx}(\omega)$ has a delta function at zero frequency due to free acceleration of the carriers. The weight of this delta function defines the charge stiffness D_c. We have

$$\Re e\sigma^{xx}(\omega) = (e^2/\hbar)D_c\delta(\hbar\omega) + \Re e\sigma_r^{xx}(\omega). \tag{8.1}$$

The free acceleration term in the real part of the frequency dependent conductivity gives rise, through the Kramers–Kronig relation, to an imaginary part which is inversely proportional to frequency, i.e.

$$\Im m\sigma^{xx}(\omega) = (2e^2/\hbar^2\omega)D_c + \Im m\sigma_r^{xx}(\omega). \tag{8.2}$$

These results can be made more concrete by taking the simplest approximation for the frequency dependent conductivity, the Drude equation,

$$\sigma(\omega) = \frac{ne^2}{m_c}\frac{\tau}{1 - i\omega\tau}$$

with a real part

$$\Re e\sigma(\omega) = \frac{ne^2}{m_c}\frac{\tau}{1 + \omega^2\tau^2}$$

and an imaginary part given by

$$\Im m\sigma(\omega) = \frac{ne^2}{m_c}\frac{\omega\tau^2}{\omega^2\tau^2 + 1}.$$

For a perfect conductor, i.e. in the limit that $\tau \rightarrow \infty$, we obtain

$$\Re e\sigma(\omega) = \frac{ne^2\pi}{m_c}\delta(\omega)$$

$$\Im m\sigma(\omega) = \frac{ne^2}{m_c\omega}$$

which leads to

$$\lim_{\omega\to 0}\omega\Im m\sigma(\omega) = \frac{ne^2}{m_c}. \tag{8.3}$$

Comparing this equation with the previous one for the charge stiffness, Eqs. 8.1 and 8.2, we find the relation between this quantity and the effective mass, namely, $D_c = n\pi \hbar^2/m_c$. Equation 8.3 was used by Kohn (1964) to discuss the metal–insulator transition due to correlations. The basic idea is that for an interacting electronic system the vanishing of D_c, or the divergence of m_c, should signal the metal–insulator transition. There is a difficulty in implementing this idea for Fermi liquids since, for an interacting Fermi-liquid the conductivity or optical mass m_c never gets renormalised by interactions due to Galilean invariance. However, this is not the case for *interacting lattice systems* as noticed by Shastry and Sutherland (1990). For these lattice systems a divergence of the optical mass is expected to occur at the localisation transition. These authors pointed out an important connection between the charge stiffness and the sensitivity of the electronic system to a change in boundary conditions. They considered the difference in ground state energies of a d-dimensional interacting fermionic lattice system, of finite size L, under a twist ϕ in the boundary conditions. They have shown that the difference in the total ground state energy density of the twisted and untwisted system can be written as (to order L^{-4})

$$\frac{E(\phi) - E(0)}{L^d} = \frac{\Delta E(\phi)}{L^d} = \frac{D_c}{L^2}\phi^2, \qquad (8.4)$$

where D_c is the charge stiffness appearing in the frequency dependent conductivity.

In their study of superconductors, Byers and Yang (1961) have shown that a change or twist in boundary conditions in a finite system of size L is formally equivalent to imagining such a system in the shape of a ring and threaded by a flux Φ. This must be accomplished in the so-called Bohm–Aharonov conditions, where the electrons in the ring are not in direct contact with the magnetic field. The flux Φ is given by the line integral of the vector potential $\mathbf{A}$ around the ring, $\oint A.dl = \Phi$. On the other hand it is related to the twist ϕ in the boundary conditions through $\Phi/\Phi_0 = \phi$ where $\Phi_0 = (hc/e)$. If we extend the Byers and Yang argument for normal and insulating rings then, according to Eq. 8.4, the insulator for which $D_c = 0$ has an additional symmetry compared to the normal perfect conductor, namely gauge invariance, since its ground state energy does not depend on the flux threading an insulating ring. This distinguishes the insulator from the perfect conductor for which the energy of the individual states and in principle the total ground state energy density depends on the flux through the ring at least to order L^{-2}.

This last assertion concerning flux sensitivity of a normal ring must be taken with care and its content leads to deep and rich consequences. The results of Byers and Yang (1961) aimed to distinguish a superconducting from a perfect conducting ring based on the sensitivity of a very large system to a flux threading it. They have shown that, for a ring of superconducting material, the total ground state energy depends on the enclosed flux, being in fact a periodic function of Φ with period

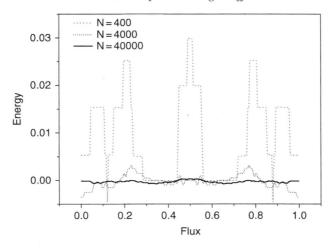

Figure 8.1 The total energy of a two-dimensional free electron gas in a ring as a function of the flux threading the ring for different systems sizes. As $N \to \infty$ the flux dependence disappears as the energy curve becomes a flat line.

$\Phi_0/2$. This is not the case, however, for a perfect conductor, which is not strictly one-dimensional. For this system, although the individual energy levels depend on the flux, the total energy does not. The way the threading flux cancels out from the expression for the total energy of a perfect conductor is a macroscopic effect and is illustrated in Fig. 8.1 for systems of different sizes. The ground state energy of a normal one-dimensional ring occurs for $\Phi = 0$ or $\Phi = \Phi_0/2$ (modulo 1) depending on whether the total number of particles N in the ring is odd or even, respectively. A flat 2d ring or a three-dimensional system can be decomposed in many $1d$ systems where the number of particles fluctuates. In a macroscopic system these fluctuations cancel out the flux dependence of the total energy as shown in Fig. 8.1. For small, mesoscopic systems, flux dependence does occur, leading to interesting and important behaviour (Büttiker *et al.*, 1983).

How do we reconcile the Byers and Yang criterion for superconductivity with the results of Shastry and Sutherland relating the charge stiffness to the flux or boundary condition dependence of the ground state energy of a *perfect conductor*? The point is that the result given by Eq. 8.4 is valid only for small values of ϕ that is for $\phi < \phi_c \sim 1/L^{d-1}$ (Scalapino *et al.*, 1993). For larger values of ϕ level crossing occurs giving rise in the limit of large systems and for $d > 1$ to the cancellation effects which provides the basis for the Byers and Yang criterion to distinguish a perfect conductor from a superconductor.

This situation is not satisfactory, especially in computer simulations where in general the size of the systems are not very large. Another criterion to distinguish metals from superconductors and insulators in a lattice has been proposed by

Scalapino *et al.* (1992). These authors considered lattice models with the kinetic energy described by

$$K = -t \sum_{ijs} (c_{is}^{\dagger} c_{js} + c_{js}^{\dagger} c_{is}),$$

where $c_{is}^{\dagger}$ and c_{is} are creation and annihilation operators respectively. They examined the current response to a space and time dependent vector potential $A_x(r, t) = A_x(q)e^{i(q \cdot r - \omega t)}$ and obtained that the uniform ($q = 0$) frequency dependent conductivity is given by

$$\sigma_{xx}(\omega) = -\frac{<k_x> - \Gamma_{xx}(q = 0, \omega)}{i(\omega + i\delta)}, \tag{8.5}$$

where the linear current response function is determined by Kubo's formula:

$$\Gamma_{xx}(q, \omega) = \frac{1}{N} \int_0^{\beta} d\tau e^{i\omega_m \tau} < j_x^p(q, \tau) j_x^p(-q, 0) > . \tag{8.6}$$

The paramagnetic current density given by

$$j_x^p(l) = it \sum_s (c_{(l+x)s}^{\dagger} c_{ls} - c_{ls}^{\dagger} c_{(l+x)s})$$

and the local kinetic energy by

$$< k_x > = -t < \sum_s (c_{(l+x)s}^{\dagger} c_{ls} + c_{ls}^{\dagger} c_{(l+x)s}) > .$$

The charge stiffness D_c can be obtained from these equations

$$\frac{D_c}{\pi e^2} = < -k_x > - \Gamma_{xx}(q = 0, \omega \to 0). \tag{8.7}$$

It is also convenient to define a *superfluid weight* D_s through the current response to a transverse, *static* gauge potential $A_x(q_y)$,

$$j_x(q_y \to 0) = -\frac{D_s e^2}{c} A_x(q_y \to 0),$$

which is London's equation and leads to the Meissner effect. The superfluid weight is related to the superfluid density ρ_s by $D_s = \rho_s/m$ where m is the mass of the superconducting carriers. The superfluid weight is given in terms of the current response function by

$$\frac{D_s}{\pi e^2} = < -k_x > - \Gamma_{xx}(q_x = 0, q_y \to 0, \omega = 0). \tag{8.8}$$

Notice that the basic difference between Eqs. 8.7 and 8.8 is the order in which the limits $q \to 0$ and $\omega \to 0$ are taken. Finally, the different ground states can be distinguished by

$$Insulator \ D_c = 0 \ D_s = 0$$
$$Metal \ D_c \neq 0 \ D_s = 0$$
$$Superconductor \ D_c \neq 0 \ D_s \neq 0$$

with D_c and D_s given by Eqs. 8.7 and 8.8 respectively.

Finally, we point out that an equation similar to Eq. 8.4 holds for a superfluid, with the superfluid weight D_s replacing the charge stiffness D_c. For superfluids this equation is valid for arbitrary values of ϕ and together with finite size scaling, they are very useful since they allow the extraction of D_s or the superfluid density in numerical studies.

8.2 Scaling Properties Close to a Metal–Insulator Transition

Charge Stiffness and Conductivity Mass

Under the assumption that the zero-temperature metal–insulator transition is a continuous, second-order phase transition, the scaling properties of the charge stiffness D_c close to this transition can be obtained. We can use for this purpose either finite size scaling theory together with Eq. 8.4 or more directly the scaling form of the frequency-dependent conductivity $\sigma(\omega)$. Let us consider the latter approach. On dimensional grounds we can write for the scaling form of the frequency-dependent conductivity,

$$\sigma(\omega) = (e^2/\hbar)\xi^{2-d} f(\omega\tau_\xi) \tag{8.9}$$

where $\tau_\xi = \xi^z$ and the correlation length, $\xi = |g|^{-\nu}$, can be identified in the metallic phase as a characteristic screening length. The quantity g as in previous chapters measures the distance of the system in parameter space to the quantum critical point associated with the metal–insulator transition at $g = 0$. The exponents ν and z are the correlation length and the dynamical critical exponents that were introduced in previous chapters and are characteristic of the quantum critical point. In the derivation of the scaling behaviour of the charge stiffness using the above scaling form of the conductivity, we make use of the asymptotic properties of the scaling function $f(\omega\tau_\xi)$ in Eq. 8.9 at certain limits. Specifically, for a perfect conductor $\Im m f(\omega\tau_\xi \to 0) \propto 1/\omega\tau_\xi$ (see Eq. 8.3) which together with the expression $\Im m \sigma(\omega) = ne^2/m_c\omega$ yield for the scaling behaviour of the charge stiffness, $D_c \propto 1/m_c$, close to the metal-insulator transition,

$$D_c \propto 1/m_c \propto \xi^{-(d+z-2)} \tag{8.10}$$

or, using hyperscaling,

$$D_c \propto |g|^{2-\alpha-2\nu} . \tag{8.11}$$

The critical exponents α, ν and the dynamical exponent z are associated with the unstable zero-temperature fixed point controlling the transition.

It is straightforward to show that the superfluid weight D_s in Eq. 8.8 and consequently the superfluid density ρ_s of a superfluid close to a $T = 0$ superfluid–insulator transition scale just like the charge stiffness D_c of the perfect metal, as in Eq. 8.10 (Fisher *et al.*, 1989).

Notice that for the metal–insulator transition that has not a well-defined order parameter, the quantum hyperscaling relation $2 - \alpha = \nu(d + z)$ plays a crucial role since it involves all the critical exponents that characterise the phase transition at zero and finite temperatures. For the superfluid, one can consider other independent exponents, as β for the order parameter, since in this case this is a well-defined quantity.

Thermal Mass

As in the previous study of the heavy fermion problem, we may also identify in the case of a metal–insulator transition a new energy scale, $T_{coh} \propto |g|^{\nu z}$. In the metallic side of the phase diagram it is natural to associate this characteristic temperature, with the coherence temperature which marks the onset of Fermi liquid behaviour with decreasing temperature, as we did for heavy fermion systems. Below this coherence line we find then a *highly correlated electron system* which is a Fermi liquid with parameters renormalised due to the proximity of the metal–insulator transition. The scaling properties of the thermal mass m_T which is proportional to the coefficient of the linear term of the specific heat can be easily obtained as in previous chapters. We find

$$m_T \propto |g|^{2-\alpha-2\nu z} \propto |g|^{\nu(d-z)}.$$

The thermal mass scales differently from the conductivity mass m_c, Eq. 8.10, obtained before. In the next two chapters we will study specific types of metal–insulator transitions.

8.3 Different Types of Metal–Insulator Transitions

As mentioned in Section 8.1, we consider here two types of metal–insulator transitions. Firstly those driven by varying the chemical potential, the density-driven transitions which may occur even in non-interacting systems. The second type are correlation-induced metal–insulator transitions arising from the competition between kinetic and Coulomb energies. The best known case of this transition is the metal–insulator transition as described by the Gutzwiller approximation of the half-filled Hubbard model (Hubbard, 1963, 1964a, 1964b) that will be discussed in the next chapter.

For density-driven transitions the control parameter $|g| = |\mu - \mu_c|$, where μ is the chemical potential and μ_c the critical value of the chemical potential at which the $T = 0$, MI transition takes place. The singular part of the free energy density, close to the transition can be written as

$$f_s \propto |\mu - \mu_c|^{2-\alpha}$$

which defines the critical exponent α. The compressibility is given by:

$$\kappa \propto \frac{\partial^2 f_s}{\partial \mu^2} \propto |\mu - \mu_c|^{-\alpha}. \tag{8.12}$$

The exponent α is related to the exponent ν of the diverging length and to the dynamic critical exponent z by the quantum hyperscaling relation, $2 - \alpha = \nu(d+z)$.

It is sometimes useful to change the control parameter of the transition from $|\mu - \mu_c| \rightarrow |n - n_c|$, where n is the density of carriers and n_c its critical value. These quantities are related by

$$n \propto \partial f_s / \partial \mu \propto |\mu - \mu_c|^{1-\alpha}.$$

If $(1 - \alpha) \geq 1$, in general, the relation between n and μ is regular, i.e. $|\mu - \mu_c| \propto |n - n_c|$; otherwise,

$$|\mu - \mu_c| \propto |n - n_c|^{\frac{1}{1-\alpha}}.$$

Using the relations above we can interchange between the different control parameters. We emphasise that Eq. 8.12 is valid only for density-driven MI transitions.

For the case of correlation-induced MI transitions a different type of scaling holds. The control parameter, for example, in the half-filled Hubbard model is $|g| \propto |U - U_c|$, where U_c is the critical value of the Coulomb repulsion at which the transition occurs. In this case the singular part of the free energy density scales as $f_s \propto |U - U_c|^{2-\alpha}$, which again defines the exponent α. For obtaining the scaling of the compressibility, it is necessary to find the scaling dimension of the chemical potential close to fixed point controlling the correlation- or interaction-induced transition. In general, since μ, as temperature, is a parameter, it scales as

$$\left(\frac{\mu}{U}\right)' = b^z \left(\frac{\mu}{U}\right), \tag{8.13}$$

where z is the dynamic critical exponent which governs the scaling of the interaction U at the zero-temperature fixed point associated with the correlation-induced metal–insulator transition. The equation above leads to the scaling form of the free energy

$$f_s \propto |U - U_c|^{2-\alpha} F\left[\frac{\mu/U}{|U - U_c|^{\nu z}}\right] \tag{8.14}$$

close to the interaction-driven metal–insulator transition at U_c. This equation allows the scaling behaviour of the compressibility to be obtained as a function of the distance to this transition, namely $\kappa \propto |U - U_c|^{2-\alpha-2\nu z}$.

Notice that the two types of MI transitions can take place in the same system or model. In the Gutzwiller approach to the Hubbard model, the insulating phase can be reached either by fixing the electron density at $n_c = 1$ and increasing the Coulomb interaction to its critical value U_c or by fixing $U > U_c$ and varying the electron density n.

8.4 Disorder-Driven Superfluid–Insulator Transition

In this section we present a numerical study of the superfluid–insulator quantum phase transition of a one-dimensional system of bosons in a lattice (Cestari *et al.*, 2010, 2011). The analysis of the numerical results is carried out using the scaling relations given above for the superfluid density. We treat non-interacting bosons such that the zero-temperature transition is driven exclusively by disorder. As we will see, this is a very special type of disorder since it also involves a commensurability or, better, an incommensurability condition. Although the disordered, non-interacting problem that we study is in a special universality class, the scaling laws for the physical quantities are the same for the pure interacting system.

In fermionic systems, disorder-driven metal–insulator transitions in the absence of correlations are known as *Anderson transitions* and they separate metallic from insulating phases.

The Hamiltonian describing non-interacting particles on a lattice is given by

$$H = \sum_i \varepsilon_i n_i + \Omega \sum_{\langle ij \rangle} (a_i^\dagger a_j + a_j^\dagger a_i), \tag{8.15}$$

where $a_i^\dagger$ and a_i are the creation and annihilation operators for a particle, boson or fermion, at the lattice site i, $n_i = a_i^\dagger a_i$ is the corresponding number operator. At each site, the particle has a local or on-site energy ε_i, hopping between sites is restricted to nearest neighbours, with amplitude Ω. We measure energies from now on in units of the tunnelling amplitude Ω. Disorder is introduced assuming a distribution for the on-site energies. For fermions, in the usual investigations of the Anderson transition, one assumes a distribution of on-site energies generally of the retangular type, i.e. with probability $P(\epsilon_i) = 1/W$ for ϵ_i in the interval $[0, W]$ and zero otherwise. The lower critical dimension for the Anderson transition has been determined to be $d = 2$. This means that all the states are localised in one dimension for any amount of disorder implying that there is no finite critical value of disorder (Lee and Ramakrishnan, 1985).

For the bosonic case, we take here a distribution of local energies which is not random, but periodic with a period incommensurate with the lattice spacing. This is generally known as the Aubry–André model (Aubry and André, 1980). The on-site energies are given by

$$\varepsilon_i = \Delta \cos(2\pi\beta i), \tag{8.16}$$

where $\beta = (1+\sqrt{5})/2$ is the golden ratio, and i assumes integer values from 1 to L. This is actually a special case of the Harper model (Harper, 1955) for electrons in a two-dimensional lattice in the presence of a perpendicular magnetic field, for which Eq. (8.16) holds for any value of β, with different characteristics of the spectrum for rational or irrational values. Disorder-like effects here are a consequence of the incommensurability between the *external potential* and the lattice. Aubry and André proved that for this model localisation only occurs when the strength of the potential Δ is larger than the critical value $\Delta_c = 2$.

For finite lattices, it is convenient to replace β with $\beta_n = F_{n+1}/F_n$, the ratio of two consecutive Fibonacci numbers, whose limit for $n \to \infty$ is the golden ratio. Then, the lattice size must be chosen as $L = F_n$ in order to allow for the use of periodic boundary conditions. For this kind of finite lattices the critical value $\Delta_c = 2$ remains a rigorous result (Ingold *et al.*, 2002).

The analysis of the numerical results is based on the scaling relation for the singular part of the superfluid density ρ_s close to the quantum superfluid-insulator phase transition. This quantity as we saw previously scales like Eq. 8.10 for the charge stiffness. We have

$$\rho_s \sim \xi^{-(d+z-2)} \sim |g|^{\nu(d+z-2)}, \tag{8.17}$$

where g measures the distance to the quantum critical point (QCP), ν is the correlation length exponent (i.e. the correlation length diverges as $\xi \sim |g|^{-\nu}$ at the QCP), d the spatial dimension, and z the dynamic critical exponent associated with the QCP. The superfluid density can be viewed as a measure of the system response to a phase-twisting field as in Eq. 8.4. It is thus natural to interpret the correlation length as a phase-coherence length. In the insulating phase it should coincide with the localisation length, which measures the spatial extent of the wave functions. This holds also for disordered metals (Lee and Ramakrishnan, 1985).

In a finite system even at criticality the correlation length is limited by the system size L, the finite-size-scaling form of the superfluid density is

$$\rho_s \sim L^{-(d+z-2)} F(L/\xi) = L^{-(d+z-2)} F(L|g|^{\nu}). \qquad (8.18)$$

The corresponding relation for the superfluid fraction ($f_s = L^d \rho_s$) is

$$f_s \sim L^{-(z-2)} F(L|g|^{\nu}). \qquad (8.19)$$

This last equation is suitable to determine the critical exponents ν and z from a numerical evaluation of f_s for various lattice sizes.

The superfluid fraction was found to undergo a very sharp transition around $\Delta = 2$ for all lattice sizes (Cestari *et al.*, 2010). This sharpness makes it difficult to directly extract the correlation length exponent from a unique although large system size. The most convenient way to obtain the critical exponents is to concentrate in a narrow region around Δ_c. Guided by the scaling form of the superfluid fraction, Eq. 8.19 and using $g \equiv \Delta - \Delta_c = \Delta - 2$, the overlap of the data for different lengths determines the critical exponents. The results are shown in Fig. 8.2. The data collapse onto two universal curves, for even and odd numbers of lattice sites. Although the scaling functions are different for these two cases, the critical exponents for which the curves collapse are the same. In view of Eq. (8.19) it is possible to identify the correlation-length exponent $\nu = 1$ from the x-axis scaling variable in Fig. 8.2, and the dynamic exponent $z = 2.374$ from the y-axis scaling. The scaling confirms the value $\nu = 1$ (Aubry and André, 1980) for this model.

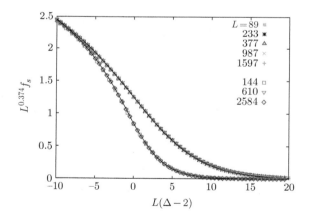

Figure 8.2 Finite size scaling of the superfluid fraction for the Aubry–André model. The data collapse in different curves for even and odd numbers of lattice sites (respectively, lower and upper curves). Reprinted figure with permission from Cestari *et al.*, (2011), Physical Review A 84, 055601. Copyright 2011 by the American Physical Society.

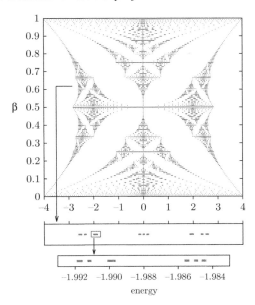

Figure 8.3 Spectra of the Harper model, highlighting the spectrum corresponding to the Aubry–André model (for a rational approximant of the golden ratio $\beta = 987/610$). Its fractal nature is illustrated in the bottom by expanding the small box drawn inside the middle panel. We actually show the spectrum for $(\beta - 1)$, which is the same as for β, according to Eq. (8.16). Reprinted figure with permission from Cestari *et al.*, (2011), Physical Review A 84, 055601. Copyright 2011 by the American Physical Society.

The value obtained for the dynamic exponent z is rather unconventional but can be understood in light of the properties of the energy spectrum of the model.

The spectrum of the Harper model has been thoroughly studied in the past (Kohmoto, 1983; Hiramoto and Kohmoto, 1989). For general rational values of β it is multifractal at $\Delta = 2$, yielding the famous *Hofstadter butterfly* (Hofstadter, 1976), shown in Fig. 8.3. There is highlighted the case treated here, with β being a rational approximant of the golden ratio. In particular, the figure shows results obtained from numerical diagonalisation of the Hamiltonian for a lattice of size $L = 610$. The two bottom panels illustrate the fractal nature of this spectrum.

With the replacement of β by a ratio of two Fibonacci numbers, $\beta_n = F_{n+1}/F_n$, the spectrum is equivalent to the one for $\bar{\beta}_n = \beta_n - 1 = F_{n-1}/F_n$, which contains F_n bands and F_{n-1} gaps. When $F_n = L \to \infty$ the width ΔE_L of a given band belonging to the spectrum scales as $\Delta E_L \sim L^{-\gamma}$, with different regions of the spectrum associated with different values of γ (not to be confused with the susceptibility critical exponent). In particular, a maximum value $\gamma_{max} = 2.374$ corresponds to band-edge states. On the other hand, the band width is a characteristic energy of the system, and therefore should scale as ξ^{-z}, which means that $\Delta E_L \sim L^{-z}$. The result $z = \gamma_{max}$ is in agreement with the relevant state for the zero-temperature superfluid–insulator transition being the bottom edge of the lowest-lying band.

In this chapter we have studied the scaling properties of the physical quantities that characterise metal–insulator and superfluid–insulator transitions. In the former case this is the charge stiffness and in the latter the superfluid density. The exponents that govern the scaling of these quantities are the correlation length exponent ν, the dynamic exponent z and the dimensionality of the system d. These are related by the quantum hyperscaling relation, $2 - \alpha = \nu(d + z)$, where the *thermodynamic* α controls the behaviour of the singular part of the free energy close to the quantum critical point. For metal–insulator transitions, in spite of there being no well-defined order parameter, it is in general always possible to define a characteristic length that diverges at the transition with the exponent ν. We have seen an example where even for a system with a complex, multi-fractal spectrum the dynamic exponent at the QCP can still be identified.

9

Density-Driven Metal–Insulator Transitions

9.1 The Simplest Density-Driven Transition

The simplest metal–insulator (MI) transition is that due to band filling (or emptying) in a system of non-interacting electrons. This is the fermionic equivalent of Bose–Einstein condensation in a system of non-interacting bosons. Let us consider a system of non-interacting electrons described by a tight-binding band in a hyper-cubic lattice in d-dimensional space. As the number of electrons per site n, for a given spin direction, approaches the value $n_c = 1$ the system evolves from a metallic to an insulating state. This transition is a quantum critical phenomenon with which we can associate critical exponents. We can introduce a characteristic length ξ, which can be identified with a screening length which diverges at the metal–insulator transition. The natural variable to describe this zero-temperature transition is the distance from the chemical potential or Fermi level μ, to the bottom, $E_b = 0$, or to the top of the band E_t, i.e. $g = \mu - E_{b,t}$. In terms of this variable, $\xi \propto |g|^{-\nu}$ and the ground state free energy density has a *singular* part that behaves as, $f \propto |g|^{2-\alpha}$, close to the transition. This expression defines the critical exponent α. The critical exponents ν and α associated with this zero temperature transition are related through the quantum hyperscaling relation $2 - \alpha = \nu(d + z)$.

For this simple MI transition it is possible to determine its universality class for any dimension (Continentino, 1995a). One starts by noticing that the gap for excitations is actually a linear function of the distance $g = \mu - E_t$. This fixes the gap exponent at the value $\nu z = 1$. Furthermore, for a hypercubic lattice the density of states close to the band edges varies as, $\rho(\omega) \propto \omega^{(d-2)/2}$, such that, the number of electrons per spin direction $n = \int_{E_b=0}^{\mu} \rho(\omega) d\omega = \mu^{d/2}$ where μ is the chemical potential. Notice also that at zero temperature the product μn has the same scaling dimension of the free energy density, $f = E - \mu n$, which is $(d + z)$. Consequently,

$$(\mu n)' = b^{d+z}(\mu n). \tag{9.1}$$

115

Also, we have

$$n' = b^d n$$
$$\mu' = b^z \mu$$

where the first equation is the statement that the number of electrons scales with the volume. In these equations b is the scaling factor. Using the relation $n = \mu^{d/2}$, we can immediately determine the dynamic exponent $z = 2$. Since $\nu z = 1$, $z = 2$, we find $\nu = 1/2$. Using the quantum hyperscaling relation, Eq. 1.18, we find $\alpha = 1 - d/2$, which can alternatively be determined by a straightforward calculation, in this case confirming the quantum hyperscaling relation. This band-filling transition is a special type of a *Lifshitz transition*. In the original literature (Kaganov and Möbius, 1984; Blanter *et al.*, 1994) this is also known as the 5/2-transition due to the behaviour of the free energy in $3d$-systems close to the QCP, $f_s \propto |g|^{5/2}$. This power is a direct consequence of the critical exponents obtained above and quantum hyperscaling with $d = 3$.

From the scaling expression for the singular part of the free energy density,

$$f \propto |g|^{1+d/2}$$

we can obtain the compressibility, $\kappa \propto \partial^2 f / \partial \mu^2$, which is given by $\kappa \propto |g|^{d/2-1}$. The susceptibility behaves as the density of states and is given by $\chi \propto |g|^{d/2-1}$. The thermal mass m_T, defined as the coefficient of the linear term of the specific heat, scales as $m_T \propto |g|^{\nu(d-z)}$ or $m_T \propto |g|^{d/2-1}$. Then, all the quantities above scale as the density of states close to the edges of the band.

The charge stiffness defined in the last chapter, $D_c \propto (n/m^*) \propto \xi^{-(d+z-2)}$ where n is the number of carriers per unit volume and m^* the conductivity mass. Then for density-driven transitions we find $D_c \propto \xi^{-d}$. In this case then $D_c \propto (Volume)^{-1}$ and it is more appropriate to associate the vanishing of D_c at the metal–insulator transition in this non-interacting case, with a decrease in the number of carriers per unit volume than with a diverging conductivity mass. Also using the results of the previous chapter we find that the conductivity at the transition, $g = 0$, scales with temperature as $\sigma(T) \propto T^{(d-2)/2}$. Since $d = 2$ is neither the upper nor the lower critical dimension for this kind of metal–insulator transition, using arguments similar to those of Fisher *et al.* (1990), we are led to expect a universal value for the conductivity at the band filling metal–insulator transition in $d = 2$ (Continentino, 1995a). In three dimensions the conductivity vanishes at the transition as $\sigma(T) \propto \sqrt{T}$.

As for the density–density correlation function, $C(r) = < n(r)n(0) > - < n(0) >^2$, we get (Sachdev, 1996),

$$C(r) = -\left| \frac{1}{V} \sum_{\mathbf{k}} \frac{e^{i\mathbf{k}.\mathbf{r}}}{e^{\beta(\epsilon_{\mathbf{k}} - \mu)} + 1} \right|^2 \approx \frac{f(r/\xi)}{r^{2d}}$$

for $r \sim \xi \to \infty$. The results of Chapter 1 allow us to identify, using $\nu = 1/2$, $z = 2$, that $\eta = d$. This result is different from that of Landau and Lifshitz (1980) for the $3d$ electron gas, $C(r) \propto 1/r^4$, obtained in a different limit, $r \gg \xi \propto |g|^{-1/2}$.

In the one-dimensional case the transition described above is in the universality class of the $d = 1$, $XY - model$ in a transverse magnetic field. This transition is the so-called Pokrovsky–Tapalov (1979) transition. This similarity points out the non-trivial aspects of the transition above which can however be fully characterised using the scaling approach.

Sometimes it is convenient to express the critical behaviour in terms of the number of electrons, instead of the chemical potential. For this purpose we note that $n \propto \partial f/\partial \mu$ and the relevant relation for spinless fermions is $|n - 1| \propto |g|^{1-\alpha} = |g|^{\frac{d}{2}}$ with $g = \mu - E_b$.

9.2 Renormalisation Group Approach

The results above can be obtained within a renormalisation group treatment of a one-dimensional tight-binding chain with linear spacing a (Oliveira *et al.*, 1984). The energy of the quasi-particles is given in terms of the wave vector k by,

$$\omega = 2t \cos ka \tag{9.2}$$

where t is the hopping integral. We now perform a scaling transformation in the linear chain by a scaling factor $b = 2$ as shown in Fig. 9.1. Then

$$a' = a/2 \tag{9.3}$$
$$k' = 2k \tag{9.4}$$

and the new energy in the renormalised chain is

$$\omega' = 2t \cos 2ka. \tag{9.5}$$

Using that $\cos 2ka = (\cos^2 ka - 1)/2$ in Eq. 9.5, we obtain the following renormalisation group equation for the energies of the excitations in the original and renormalised chain,

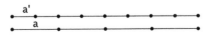

Figure 9.1 A renormalisation group transformation for a linear chain by a scaling factor $b = 2$.

$$\frac{\omega_{n-1}}{t} = \left(\frac{\omega_n}{t}\right)^2 - 2, \tag{9.6}$$

which we write as

$$\Omega_{n+1} = \Omega_n^2 - 2 \tag{9.7}$$

with $\Omega = \omega/t$. The fixed points of this equation are easily obtained and we get $\Omega_1^* = -1$ and $\Omega_2^* = 2$. Since $|\partial\Omega_{n+1}/\partial\Omega_n|_{\Omega_1^*,\Omega_2^*} \geq 1$ both fixed points are unstable. The interesting one is $\Omega_2^* = 2$, located at the band edge $\omega = 2t$.

The dynamic exponent z is given by the recursion relation, $\Omega' = b^z\Omega$ in the neighbourhood of the fixed point Ω_2^*. We get, using $b = 2$,

$$z = \frac{\ln\left|\frac{\partial\Omega_{n+1}}{\partial\Omega_n}\right|_{\Omega_2^*}}{\ln b} = 2 \tag{9.8}$$

where we used $b = 2$. This value for the dynamic exponent is the same we found previously using a scaling analysis. Equation 9.7 is a fully chaotic equation which has been much studied. Any initial point Ω_0 in the interval $[-2, 2]$ will always iterate inside this interval. As the number of iterations becomes very large, a value of the sequence will get arbitrarily close to any given point in the interval. The density of visits $n(\Omega)$ of a given interval of width $d\Omega$, actually gives the density of states $\rho(\omega)$ of the linear chain, i.e.

$$n(\Omega)d\Omega = \rho(\omega)d\omega.$$

This relation can be extended for arbitrary dimensions and different types of lattices. For the case of a body-centred cubic lattice in three dimensions, we define three sequences:

$$\begin{aligned}
\Omega_{n+1}^x &= \Omega_n^{x\,2} - 2 \\
\Omega_{n+1}^y &= \Omega_n^{y\,2} - 2 \\
\Omega_{n+1}^z &= \Omega_n^{z\,2} - 2.
\end{aligned} \tag{9.9}$$

The sequence of numbers $\Omega_{n+1} = \Omega_{n+1}^x + \Omega_{n+1}^y + \Omega_{n+1}^z$, obtained from the iteration of Eqs. 9.9 for any set of different initial values $\Omega_0^x, \Omega_0^y, \Omega_0^z$ in the interval $[-2, 2]$, will always yield a number in the range $[-6, 6]$. Counting the number of visits to a given interval in this range for a large number of iterations yields the density of states for a cubic lattice as shown in Fig. 9.2.

For a sequence of iterations, $\Omega_{n+1} = \Omega_{n+1}^x\Omega_{n+1}^y\Omega_{n+1}^z$ for any initial set $\Omega_0^x, \Omega_0^y, \Omega_0^z$ in the interval $[-2, 2]$ will always yield a number in the range $[-6, 6]$. The density of visits in this case gives the density of states of the *bcc* lattice as

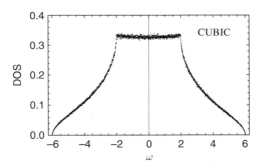

Figure 9.2   The density of states of a cubic lattice obtained from iterating Eqs. 9.9.

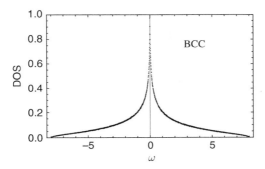

Figure 9.3 Density of states for a body-centred cubic lattice obtained from iterating Eqs. 9.9.

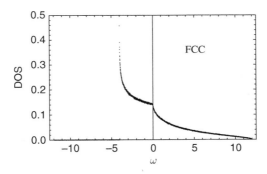

Figure 9.4 Density of states for a face-centred cubic lattice obtained from iterating Eqs. 9.9.

shown in Fig. 9.3. In the same way the sum of the products $\Omega_{n+1} = \Omega_{n+1}^x \Omega_{n+1}^y + \Omega_{n+1}^x \Omega_{n+1}^z + \Omega_{n+1}^y \Omega_{n+1}^z$ yields a number in the interval $[-4, 12]$. The density of visits of the sequence Ω_{n+1} yields the density of states of the fcc lattice in this interval shown in Fig 9.4.

9.3 Metal–Insulator Transition in Divalent Metals

What is the relevance of density-driven transitions for real systems? Certainly doping is not the most convenient way of accessing this transition since it introduces disorder and is difficult to control. Interesting candidates to investigate density-driven metal–insulator transition are divalent metals like Ytterbium or Strontium which have a semi-metal-like band structure, i.e. two wide bands of uncorrelated electrons intersecting at the Fermi level (see Fig. 9.5). In this case, by applying external pressure we can increase the hybridisation of these bands giving rise to a *repulsion* between them which has the effect of reducing the density of states in the region of overlap as shown in Fig. 9.5.

Let us study the metal–insulator transition which is driven by increasing the hybridisation among the bands in a semi-metal. We will show that this transition is in the same universality class of the density-driven metal–insulator transition studied before. For this purpose we introduce the following Hamiltonian:

$$H = \sum_{i,j,\sigma} t^1_{ij} c^{1\dagger}_{i\sigma} c^1_{j\sigma} + \sum_{i,j,\sigma} t^2_{ij} c^{2\dagger}_{i\sigma} c^2_{j\sigma} + V \sum_{i,\sigma} \left(c^{1\dagger}_{i\sigma} c^2_{i\sigma} + c^{2\dagger}_{i\sigma} c^1_{i\sigma} \right), \qquad (9.10)$$

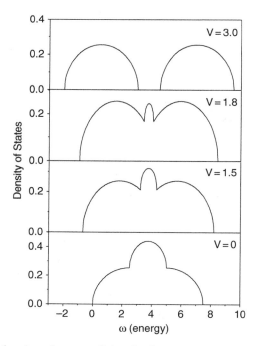

Figure 9.5 The density of states of the divalent semi-metal as the hybridisation V increases with increasing pressure driving the metal–insulator transition. The bands are always symmetric with respect to the Fermi level that remains fixed.

where t_{ij}^l gives the hopping amplitude of an electron in the lth-band ($l = 1, 2$), $c_{j\sigma}^{i\dagger}$ and $c_{j\sigma}^i$ create and destroy electrons at site j with spin σ on the lth-band, respectively. V is the hybridisation or mixing term which transfers electrons from one band to another. The dispersion relations for the bands of the semi-metal are given by: $\epsilon_k^1 = \epsilon_k$ and $\epsilon_k^2 = \alpha\epsilon_k + \Delta/2$ where Δ is a band-shift taken equal to the width of band-1. The dispersion relations are given by $\epsilon_k^l = \sum_{i,j} t_{i,j}^l exp[i\vec{k}.(\vec{R}_i - \vec{R}_j)]$. The quantity α is the ratio of the effective masses of the electrons in the two bands. In the following we consider symmetric bands with respect to the Fermi level and take $\alpha = 1$. In such a situation, for a divalent metal, the hybridised bands remain always symmetric with respect to the Fermi level that in turn stays fixed.

The Hamiltonian above can be easily diagonalised by the Green's function method. Let $\omega_1(k)$ and $\omega_2(k)$ be the new excitations of the system, which are obtained from the roots of the equation

$$(\omega - \epsilon_k^1)(\omega - \epsilon_k^2) - V^2 = 0. \tag{9.11}$$

Introducing two new functions $g_1(\omega)$ and $g_2(\omega)$ through the equation

$$[\omega - \omega_1(k)][\omega - \omega_2(k)] = [g_1(\omega) - \epsilon_k][g_2(\omega) - \epsilon_k] \tag{9.12}$$

we get, in terms of these new functions, the contributions of the ith-band to the density of states of the new hybrid bands. The band-1 contribution to the density of states of the first hybrid band is given by

$$n_{\sigma 1}^1(\omega) = \frac{|f^1[\omega, g_1(\omega)]|}{\alpha|g_1(\omega) - g_2(\omega)|} \sum_k \delta[g_1(\omega) - \epsilon_k] \tag{9.13}$$

and to the second hybrid band by

$$n_{\sigma 2}^1(\omega) = \frac{|f^1[\omega, g_2(\omega)]|}{\alpha|g_1(\omega) - g_2(\omega)|} \sum_k \delta[g_2(\omega) - \epsilon_k]. \tag{9.14}$$

For the contribution of band-2 to the first hybrid band we find

$$n_{\sigma 1}^2(\omega) = \frac{|f^2[\omega, g_1(\omega)]|}{\alpha|g_1(\omega) - g_2(\omega)|} \sum_k \delta[g_1(\omega) - \epsilon_k], \tag{9.15}$$

and for the second hybrid band

$$n_{\sigma 2}^2(\omega) = \frac{|f^2[\omega, g_2(\omega)]|}{\alpha|g_1(\omega) - g_2(\omega)|} \sum_k \delta[g_2(\omega) - \epsilon_k]. \tag{9.16}$$

The functions $f^1[\omega, \epsilon_k] = \omega - \epsilon_k^2$ and $f^2[\omega, \epsilon_k] = \omega - \epsilon_k^1$ where, as before, $\epsilon_k^1 = \epsilon_k$ and $\epsilon_k^2 = \alpha\epsilon_k + \beta$ with $\alpha = 1$ and $\beta = \Delta/2$ in the case of the divalent semi-metal. The functions $g_{1,2}(\omega)$ are given by:

$$g_{1,2}(\omega) = \omega - \frac{\Delta}{4} \pm \left[V^2 + \left(\frac{\Delta}{4} \right)^2 \right]^{1/2}. \tag{9.17}$$

The energies of the bottom of the hybrid bands are

$$E_b^{2,1} = \frac{\Delta}{4} \pm \left[V^2 + \left(\frac{\Delta}{4} \right)^2 \right]^{1/2}$$

and the energies of the top

$$E_t^{2,1} = \frac{5}{4}\Delta \pm \left[V^2 + \left(\frac{\Delta}{4} \right)^2 \right]^{1/2}.$$

For the gap between the bands we get

$$\Delta_G = E_b^2 - E_t^1 = 2 \left[V^2 + \left(\frac{\Delta}{4} \right)^2 \right]^{1/2} - \Delta. \tag{9.18}$$

A gap opens for $V > V_c$, where $V_c = \frac{\sqrt{3}}{4}\Delta$, signalling the metal–insulator transition. Furthermore the gap for V close to V_c is given by

$$\Delta_G \approx \sqrt{3}(V - V_c). \tag{9.19}$$

In the case of the divalent semi-metal, the bands are always symmetric with respect to the Fermi level which remains fixed at $\mu = \frac{3}{4}\Delta$, for any value of the hybridis-ation. In this case the distance from the top of the lower hybrid band to the Fermi level is just half the gap defined above, i.e.

$$\mu - E_t^1 = \frac{1}{2}\Delta_G. \tag{9.20}$$

For V close to V_c we find

$$\mu - E_t^1 = \frac{\sqrt{3}}{2}(V - V_c). \tag{9.21}$$

This relation allows to express the distance to the critical point at which the metal–insulator transition occurs in terms of the variable $g = \mu - E_t^1$. Furthermore, close to the transition the gap behaves in terms of this variable as given by Eq. 9.20 and from this equation we can identify the gap exponent $\nu z = 1$. It can be shown that the relevant exponents depend only on the form of the density of states of the orig-inal, unhybridised bands, close to the band edges. Then, for hypercubic lattices, we find for the hybridisation-induced metal–insulator transition, the same expo-nents governing the critical behaviour of the free energy, compressibility, etc., as for the non-interacting density-driven transition studied before. Consequently, the

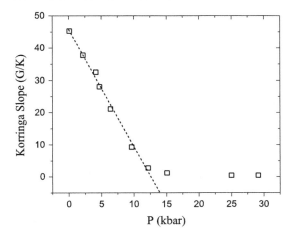

Figure 9.6 The coefficient of the Korringa linewidth of *Gd* impurities in *Yb* as a function of pressure, approaching the metal–insulator transition. This is proportional to the square of the density of states at the Fermi level or to compressibility square, $\kappa^2 \propto |g|^{2\nu(d-z)}$ (Continentino, Elschner and Jakob, 1995). The linear behaviour is associated with the critical exponents $\nu = 1/2$, $z = 2$ and $d = 3$. The quantity $g = (P - P_c)$ where the critical pressure $P_c \approx 12.5$ kbars.

metal–insulator transition which occurs due to the increase of hybridisation, for example, by applying pressure to a semi-metal like Ytterbium, provides a physical realisation of the simple density-driven transitions (see Figs. 9.5 and 9.6). The control parameter is $g = \mu - E_t^1 \propto (V - V_c)$ or still $(P - P_c)$, as in Figs. 9.5 and 9.6. Fig. 9.6 shows the linear decrease with pressure of the Korringa relaxation rate of *Gd* impurities in *Yb* (Continentino, Elschner and Jakob, 1995). For a 3d system this linear behaviour is a consequence that $\nu z = 1$ for density-driven or Lifshitz transitions (see caption of Fig. 9.6). Scaling (Enderlein, 2013) and anomalous non-Fermi liquid behaviour at the Lifshitz transition (Enderlein, 2016) has also been observed in the transport properties of elemental *Yb* under pressure.

The scaling theory can be generalised to finite temperatures and the expression for the $T \neq 0$ free energy density is given by

$$f \propto |g|^{2-\alpha} f_0 \left[T / T_c^* \right], \tag{9.22}$$

where the characteristic temperature $T_c^* \propto |g|^{\nu z}$. For $T << T_c^*$, the thermal mass, in the metallic side, $m_T \propto |g|^{\nu(d-z)}$ or using the exponents obtained before, $m_T \propto |g|^{\frac{d-2}{2}}$. Then, for $d = 3$, there is no enhancement of the thermal mass close to the MI transition.

Notice that the characteristic temperature $T_c^* \propto |V - V_c|$, ($\nu z = 1$), gives the width of the peak in the density of states at the Fermi level (see Fig. 9.5) and provides the relevant energy scale, in the metallic side, just before the metal–insulator transition. On the other hand, in the insulating side where the system is gapped T_c^* gives the energy scale for the creation of electron–hole pairs.

9.4 The Excitonic Transition

In density-driven metal–insulator transitions that occur due to the merging of two bands, one has to take into account the electron–hole interaction. The energy gain due to the formation of an electron–hole pair may overcome the energy of the gap and give rise to an excitonic phase transition (Halperin and Rice, 1968). These excitonic transitions have been the subject of intense investigations in the late sixties. Although predicted theoretically and intensively searched, these phases have never been observed in divalent metals. For density-driven transitions induced by hybridisation, as when pressure is applied in the system, we do not expect to see a true phase transition to an excitonic state. The reason is that the hybridisation V acts as a conjugate field to the order parameter characterising the excitonic phase. The effect is similar to that of a ferromagnetic system in the presence of an uniform external magnetic field which destroys the ferromagnetic transition. Consider the electron–hole interaction in its simplest form, that of a local attraction between spinless electrons and holes:

$$H_{eh} = -G \sum_i c_i^{1\dagger} c_i^1 c_i^{2\dagger} c_i^2.$$

The order parameter associated with the excitonic phase is given by $\langle c_i^{1\dagger} c_i^2 \rangle$. If we recall the hybridisation term, now for spinless particles

$$V \sum_i \left(c_i^{1\dagger} c_i^2 + c_i^{2\dagger} c_i^1 \right)$$

we can easily see that the one-body mixing term couples to the order parameter of the excitonic phase, playing the role of a conjugate field. Consequently, in the presence of hybridisation a sharp phase transition to an excitonic phase never occurs, instead we expect to observe a smooth crossover to new type of state. Although there will be strong electron–hole correlations in the presence of the interaction G, the effect of these interactions is to renormalise the bare mixing term V. The only zero temperature phase transition expected to occur in the semi-metal as pressure increases is the density-driven metal–insulator transition due to the change in hybridisation.

9.5 The Effect of Electron–Electron Interactions

The effect of electron–electron interactions in density-driven transitions in $3d$ Fermi liquids has been studied by Kaganov and Möbius (1984). These authors show that these interactions renormalise Fermi liquid parameters but do not change the universality class of the 5/2-transition. For lower dimensions the situation is more complicated since the stability of the Fermi liquid state itself is in question. Consider an effective action with a non-interacting part describing a band filling transition with $\nu = 1/2$ and $z = 2$:

$$S_{eff} = \int d^d q \int d\omega [i\omega + q^2 + (\mu - \mu_c)] |\psi(q, \omega)|^2.$$

If one adds a repulsive contact interaction with a coefficient u and remove slow modes below the Fermi surface, simple power counting shows that u renormalises as

$$u' = b^{\frac{2-d}{2}} u$$

and becomes irrelevant for $d > 2$ as expected. In the section below we use scaling arguments to investigate the nature of the density-driven metal–insulator transition in the one-dimensional Hubbard model and in a toy model. Both have exact solutions. For the former, even though a simple Fermi liquid picture is inadequate to describe its metallic phase, the density-driven transition metal–insulator transition is described by the same exponents of the non-interacting case and are related by the quantum hyperscaling relation.

9.6 The Density-Driven MI Transition in the $d = 1$ Hubbard Model

We are concerned here only with metal–insulator transitions *which are not accompanied by the appearance of long range magnetic order*. We discuss the results of two exactly soluble models and will wait to the next chapter to draw some conclusions.

The first model is the one-dimensional Hubbard model. In this case at $T = 0$, for a half-filled band, $n = 1$, the model always yields a paramagnetic insulator for any value of the Coulomb repulsion U. There is no interaction-driven MI transition in this case since $U_c = 0$. For any other band-filing, however, this model has as its ground state a metallic paramagnet. Then, for a given U, there is a density-driven metal–insulator transition, as the band-filling or chemical potential is varied. Furthermore this transition is not accompanied by the appearance of magnetic order as a consequence of the separation of charge and spin degrees of freedom at $d = 1$.

We rely on exact results obtained from the Bethe ansatz solution of this model (Kawakami and Yang, 1990), to carry on the following analysis. Firstly, we point out that the compressibility κ diverges as, $\kappa \propto (n - n_c)^{-1}$ and that the charge stiffness D_c vanishes linearly with the number of holes, i.e. $D_c \propto (n - n_c)$ at this transition where $n_c = 1$. The optical mass $m^* \propto (1/D_c) \propto (n - n_c)^{-1}$

and consequently diverges with the same numerical exponent of the compressibility. The above exact result for the critical behaviour of the compressibility, together with a rigorous relation between the singular part of m^* and this quantity, namely, $m^* \propto \kappa$, allow to determine unambiguously the critical exponents characterising this density-driven transition when expressed in terms of $g = \mu - \mu_c(U)$, where $\mu_c(U)$ is the value of the chemical potential at the phase boundary between the metal and insulator for a given U. For this purpose we notice that, $n \propto \partial f / \partial \mu \propto \mid g \mid^{1-\alpha}$. If, $(1 - \alpha) \geq 1$, i.e., $\alpha \leq 0$, the relation between μ and n is regular, i.e. $g = [\mu - \mu_c(U)] \propto (n - n_c)$, otherwise, $(n - n_c) \propto \mid g \mid^{1-\alpha}$. The compressibility, $\kappa = \partial^2 f / \partial \mu^2 \propto \mid g \mid^{-\alpha}$. Let us assume that the regular term dominates, i.e. $1 - \alpha \geq 1 (\alpha \leq 0)$. Then $(n - n_c) \propto g$ and consequently, $\kappa \propto (n - n_c)^{-\alpha}$. Due to the exact result, $\kappa \propto (n - n_c)^{-1}$, we should then have $\alpha = 1$. This is in contradiction with the initial assumption that the relation between μ and n is regular, i.e. that $\alpha \leq 0$. The alternative possibility yields $[\mu - \mu_c(u)] \propto (n - n_c)^{\frac{1}{1-\alpha}}$ and $\kappa \propto (n - n_c)^{-(\alpha/1-\alpha)}$. Comparing with the exact result, $\kappa \propto (n - n_c)^{-1}$, we get $\alpha = 1/2$. On the other hand, the optical mass, $m^* \propto \mid \mu - \mu_c(U) \mid^{-(2-\alpha-2\nu)}$, as derived before and since, $m^* \propto \kappa$, we have, $2 - \alpha - 2\nu = \alpha$ or $\nu = 1 - \alpha = 1/2$. Finally, from the hyperscaling relation, we get $z = 2$ and consequently $\nu z = 1$. We can easily check that these exponents, $\alpha = 1/2$, $\nu = 1/2$ and $z = 2$ are the same for the non-interacting, one-dimensional, density-driven transition. Consequently the density-driven MI transition in the $d = 1$ Hubbard model is in the same universality class of the simple band-filling transition of the $d = 1$ tight-binding chain.

We next present results on an interacting model, which although unrealistic, can be solved exactly and as such allows for a further comparison with the scaling results in the interacting case (Continentino and Coutinho-Filho, 1994b; Vitoriano *et al.*, 2000). The model has infinite range interactions and is described by the following Hamiltonian:

$$H = \sum_k \epsilon_k c_{k\sigma}^\dagger c_{k\sigma} + U \sum_k n_{k\uparrow} n_{k\downarrow}. \tag{9.23}$$

The phase diagram of the model for an arbitrary dimension is shown in Fig. 9.7. This model presents the two types of metal–insulator transitions discussed previously. The $n = 1$, paramagnetic, Mott insulating phase can be reached either at a constant density, $n = 1$, for $U > U_c$ or for a fixed $U > U_c$ varying the electron density or chemical potential. In the former case the chemical potential is fixed at $\mu = U/2$. The Mott phase satisfies a commensurability criterion and is incompressible since the number of electrons per site remains fixed

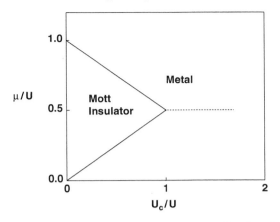

Figure 9.7 The phase diagram associated with the Hamiltonian, Eq. 9.23 (Continentino and Coutinho-Filho, 1994b).

while the chemical potential varies. Consequently $\kappa = \partial n/\partial \mu = 0$ in the Mott phase.

For this model, both the density-driven and interaction-driven transitions are in the same universality class. It turns out that, independently of the way the insulating phase is reached, the critical exponents are those of the non-interacting band-filling metal–insulator transition investigated previously, i.e. $\nu = 1/2$, $z = 2$ for any dimension and the exponent α, which depends on dimensionality, is determined by the generalised hyperscaling relation. Then, the model has no upper critical dimension, both the fixed-density and the density-driven transitions. As for non-interacting density-driven metal–insulator transitions, the exponent α always depends on dimensionality and never sticks to a mean field, d-independent value.

The results above for the $d = 1$ Hubbard model and for the model with infinite range interactions suggest that interactions do not change the universality class of the density-driven metal–insulator transition, as long as, this transition is not accompanied by the appearance of long-range magnetic order. In the next chapter we will see this is not really the case.

9.7 Effects of Disorder

How does disorder affects the results we have obtained? For $d \leq 2$ we expect drastic effects as all the states become localised and most probably the density-driven transition is in the universality class of the Anderson transition. For $d > 2$ the effect should depend on the strength of disorder, i.e. if it will give rise to localised states or not since these states appear just close to the band edges which is the relevant region of the density of states for the density-driven transition. In case there is a mobility edge the density-driven is an Anderson transition. The role of interactions in this case is subject of intense investigations (Lee and Ramakrishnan, 1985; Belitz and Kirkpatrick, 1994; Imada, Fujimori and Tokura, 1998; Georges *et al.*, 1996).

10

Mott Transitions

10.1 Introduction

In this chapter we study metal–insulator transitions which occur due to correlations between the charge carriers. This quantum phase transitions results from the competition between kinetic energy and Coulomb repulsion where the latter has the tendency to localise the electrons. The simplest model which describes this competition is the one-band Hubbard model with its two energy scales, the bandwidth W and the Coulomb repulsion U between electrons of opposite spins on the same site. For a fixed density of electrons, $n = 1$; as the ratio U/W increases, we expect to find a zero-temperature phase transition from a metallic state to an insulator where the electrons are frozen at the lattice sites. The method employed here to investigate this transition is the Gutzwiller approach (Gutzwiller, 1965). This is a variational method which played an important role in providing a framework for the ideas of Mott on the problem of the metal–insulator transition (Mott, 1974). On the metallic side of the transition, this approach describes a metal with renormalised Fermi liquid parameters, essentially with enhanced effective mass and uniform susceptibility. This is the so-called *highly correlated electron gas* (Brinkman and Rice, 1970) for which the metal oxide V_2O_3 is the prototype system. In the insulating side, Gutzwiller's approximation yields a non-interacting system of localised electrons. This insulating phase is *incompressible*, i.e. has zero compressibility and infinite zero temperature susceptibility. The latter is not a consequence of interactions but simply due to the divergence of the Curie susceptibility, $\chi \propto 1/T$, of non-interacting moments at $T = 0$. Then, in Gutzwiller's approach the transition is from a paramagnetic metal to a paramagnetic insulator. We will continue using the designation *Mott transition* to describe this type of phase transition where the insulating phase has no long-range magnetic order. In real systems, most frequently, localisation due to interactions is accompanied by the appearance of antiferromagnetic order. It should be clear that the

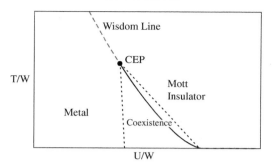

Figure 10.1 Phase diagram (schematic) of the Hubbard model within the Dynamic Mean Field Theory (DMFT). The solid line is a line of first-order transitions that terminates at a critical end point (CEP) above which there is a crossover line, the Widom line. The dotted lines mark the limit of stability of the metallic and insulating phases (Vollhardt, 2012).

Gutzwiller approach as presented here does not describe this phenomenon and is limited to the charge fluctuation aspects of the transition. The type of phenomena we treat here can also be approached using the *slave-boson* method of Kotliar and Ruckenstein (1986) and Lavagna (1990). Whenever this method yields analytical results, they always coincide with those of the Gutzwiller approximation. We therefore opted to describe in some detail the original theory, as developed by Nozières (1986), in some detail. Our goal is to look at the MI transition problem from the point of view of the theory of quantum critical phenomenon. We want to identify the relevant critical exponents governing the critical behaviour of the physical quantities close to the MI transition within Gutzwiller's approximation.

More recently, studies based on infinite-dimensional models allowed the formulation of a dynamic mean field theory (DMFT) (Georges *et al.*, 1996; Miranda *et al.*, 2008; Vollhardt, 2012; Vučičević *et al.*, 2015) that unlike Gutzwiller's approximation (GW) is not static but dynamic in nature. The character of different phases that in GW are specified by a number, like the average double occupancy of a site, are accomplished in DMFT by a function of frequency as the self-energy, which in this approximation is strictly local. This new approach incorporates features of the GW approximation, as the quasi-particle peak but also those of the so-called Hubbard III solution, the lower and upper Hubbard bands.

For the half-filling case, the resulting zero-temperature localisation transition is discontinuous. As temperature increases there is a line of first-order transitions that terminates at a critical end point at T_c. Neighbouring the first-order line there is a finite region of coexistence for a range of temperatures and of the ratio U/W. Above the critical end point (CEP), there is a *Widom line* that defines a region of crossover between the different phases. The discontinuous nature of the $T = 0$ transition restrains the application of a quantum scaling approach to this problem.

For this reason we will not pursue further the study of the DMFT here. Notice, however, that if one enlarges the phase diagram of the Hubbard model, allowing for frustration or deviations from half-filling, it is possible to bring T_c smoothly to zero and recover quantum scaling behaviour (Vučičević *et al.*, 2015).

10.2 Gutzwiller Approach

Let us consider a lattice of L sites, with $N_\uparrow$ spins up, $N_\downarrow$ spins down and D doubly occupied sites. In the absence of electronic correlations ($U = 0$), the number of doubly occupied sites is $D_0 = (n_\downarrow n_\uparrow)L$, where $n_\sigma = N_\sigma/L$. In the presence of correlations the number of doubly occupied sites must decrease since double occupancy increases the energy of the system, so we must have $D < D_0$.

Let $|\psi_0\rangle$ be the wave function of the system without interactions. The Gutzwiller wave function is given by

$$|\psi\rangle = \prod_i [1 - (1 - \varrho)n_{i\downarrow}n_{i\uparrow}]|\psi_0\rangle,$$

where ϱ is a variational parameter to be determined, such that the ground state energy of the system $E = \langle\psi|H|\psi\rangle/\langle\psi|\psi\rangle$ is a minimum. This wave function can also be written as

$$|\psi\rangle = \exp\left\{\sum_i \ln[1 - (1 - \varrho)n_{i\downarrow}n_{i\uparrow}]\right\}|\psi_0\rangle.$$

Note that

$$\ln[1 - (1 - \varrho)n_{i\downarrow}n_{i\uparrow}] = \begin{cases} 0 & \text{if } n_{i\downarrow}n_{i\uparrow} = 0 \\ \ln\varrho & \text{if } n_{i\downarrow}n_{i\uparrow} = 1 \end{cases}.$$

Consequently, $\sum_i \ln[1 - (1 - \varrho)n_{i\downarrow}n_{i\uparrow}] = (\sum_i n_{i\downarrow}n_{i\uparrow}) \ln\varrho = \hat{D}\ln\varrho$, with $\hat{D} = \sum_i n_{i\downarrow}n_{i\uparrow}$ such that the wave function can also be written as $|\psi\rangle = \varrho^{\hat{D}}|\psi_0\rangle$. Note that $\varrho = 1$ corresponds to the system without interactions. In the thermodynamic limit $|\psi\rangle$ is an eigenfunction of the operator $\hat{D}$, i.e. $\hat{D}|\psi\rangle = D|\psi\rangle$. In fact $\hat{D}|\psi\rangle = \varrho(\partial/\partial\varrho)|\psi\rangle = (d_1 + d_2 + d_3 + \cdots)|\psi\rangle = D|\psi\rangle$. As a consequence, the expectation value of the correlation term can be easily obtained in the basis $|\psi\rangle$ and we have

$$\frac{\langle\psi|U\sum_i n_{i\downarrow}n_{i\uparrow}|\psi\rangle}{\langle\psi|\psi\rangle} = UD.$$

The difficulty in this approach is then to calculate the norm $\langle\psi|\psi\rangle$ and the kinetic energy term. For this purpose we need to consider the combinatorial problem of the number of ways, $N_D(L, N_\uparrow, N_\downarrow)$, that in a lattice with L sites and $N_\uparrow$ ($N_\downarrow$) spins up (down), we find D doubly occupied sites. In the absence of correlations this is given by

$$N_D(L, N_\uparrow, N_\downarrow) = \frac{L!}{(N_\uparrow - D)!(N_\downarrow - D)!D!(L - N_\uparrow - N_\downarrow + D)!}$$

where $(L - N_\uparrow - N_\downarrow + D) = (L - (N_\uparrow - D) - (N_\downarrow - D) - D)$ is the number of empty sites. Let $P(L, N_\sigma)$ be the probability of a given configuration with N_σ spins to occur, then

$$P(L, N_\sigma) = n_\sigma^{N_\sigma}(1 - n_\sigma)^{L - N_\sigma}$$

since we have neglected spatial correlations. Using this notation we find

$$\langle \psi | \psi \rangle = \sum_D \varrho^{2D} N_D(L, N_\uparrow, N_\downarrow) P(L, N_\uparrow) P(L, N_\downarrow),$$

since $\langle \psi | \psi \rangle = \sum_D \varrho^{2D} \langle \psi_0 | \psi_0 \rangle = \sum_D \varrho^{2D} |\psi_0|^2$. For the correlation term we get

$$\langle \psi | U \sum_i n_{i\downarrow} n_{i\uparrow} | \psi \rangle = L \sum_D \varrho^{2D+2} N_D(L-1, N_\uparrow-1, N_\downarrow-1) P(L, N_\uparrow) P(L, N_\downarrow).$$

The arguments of N_D, namely, $L-1$, $N_\uparrow-1$, $N_\downarrow-1$, arise since we are considering a fixed configuration with site i doubly occupied, such that there are only $L - 1$ sites available and $N_\uparrow - 1$, $N_\downarrow - 1$ spins $\uparrow$ and $\downarrow$ to permute. The factor ϱ^{2D+2} arises since the number of doubly occupied sites is $D + 1$ in the initial and final states giving the factor $\varrho^{D+1} \cdot \varrho^{D+1} = \varrho^{2D+2}$.

The problem of summing over all configurations D above is very complicated; however, in the thermodynamic limit we may consider just the dominant contribution to this sum. Let us calculate the norm first:

$$\langle \psi | \psi \rangle = P(L, N_\uparrow) P(L, N_\downarrow) \sum_D \varrho^{2D} N_D(L, N_\uparrow, N_\downarrow).$$

For the purpose of determining the dominant term, we find the extremum of $\ln[\varrho^{2D} N_D]$ instead of $\varrho^{2D} N_D$. Defining $S = \ln[\varrho^{2D} N_D]$, using Stirling's approximation to expand N_D, we obtain from $(\partial S / \partial D)_{D=\tilde{D}} = 0$, the following relation between ϱ and $\tilde{D}$:

$$\varrho^2 = \frac{\tilde{d}(1 - n_\uparrow - n_\downarrow + \tilde{d})}{(n_\downarrow - \tilde{d})(n_\uparrow - \tilde{d})} \tag{10.1}$$

where $\tilde{d} = \tilde{D}/L$. This equation yields the $\tilde{d}(\varrho)$ which gives the maximum contribution to the sum $\sum_D$. Then, replacing this sum by this dominant term, the Coulomb repulsion part of the Hamiltonian is calculated as

$$\frac{\langle\psi|\sum_i n_{i\downarrow}n_{i\uparrow}|\psi\rangle}{\langle\psi|\psi\rangle} = \frac{L\varrho^{2\tilde{D}+2}(L-1)!(N_\uparrow-\tilde{D})!(N_\downarrow-\tilde{D})!}{(N_\uparrow-1-\tilde{D})!(N_\downarrow-1-\tilde{D})!\tilde{D}!}$$

$$\times\frac{\tilde{D}!(L-N_\uparrow-N_\downarrow+\tilde{D})!}{(L-1-N_\uparrow+1-N_\downarrow+1+\tilde{D})!\varrho^{2\tilde{D}}L!}$$

$$= \varrho^2\frac{(N_\uparrow-\tilde{D})(N_\downarrow-\tilde{D})}{L(L-N_\uparrow-N_\downarrow+\tilde{D}+1)}$$

$$= \varrho^2\frac{(n_\uparrow-\tilde{d})(n_\downarrow-\tilde{d})}{(1-n_\uparrow-n_\downarrow+\tilde{d})}.$$

Using the value of ϱ^2 obtained before, we finally get

$$\frac{\langle\psi|U\sum_i n_{i\downarrow}n_{i\uparrow}|\psi\rangle}{\langle\psi|\psi\rangle} = U\tilde{d}.$$

The Kinetic Energy Term

We want now to calculate the kinetic energy term

$$K = \frac{\langle\psi|\sum_{ij} c_{i\uparrow}^\dagger c_{j\uparrow}|\psi\rangle}{\langle\psi|\psi\rangle}.$$

For this purpose we need to consider the four processes by which a spin $\uparrow$ electron can jump from site j to site i. Let us consider these processes.

First Gutzwiller's Hopping

$$\uparrow j \bullet - - - - - \bullet i \Longrightarrow j \bullet - - - - - - \bullet i \uparrow$$

$$\langle\psi|c_{i\uparrow}^\dagger c_{j\uparrow}|\psi\rangle = \sum_D \varrho^{2D}N_D(L-2,N_\uparrow-1,N_\downarrow)P(L-2,N_\uparrow-1)P(L,N_\downarrow)$$

Notice the value of the argument $L-2$ in N_D since we have excluded two sites (i and j) from the total number of sites. Also, since we fixed one spin $\uparrow$, we get $N_\uparrow-1$ in N_D. We then have

$$\frac{\langle\psi|c_{i\uparrow}^\dagger c_{j\uparrow}|\psi\rangle}{\langle\psi|\psi\rangle} = \frac{\varrho^{2\tilde{D}}(L-2)!(N_\uparrow-\tilde{D})!(N_\downarrow-\tilde{D})!(L-N_\uparrow-N_\downarrow+\tilde{D})!}{(N_\uparrow-1-\tilde{D})!(N_\downarrow-\tilde{D})!(L-N_\uparrow-N_\downarrow+\tilde{D}-1)!}$$

$$\times\frac{\tilde{D}!n_\uparrow^{N_\uparrow-1}(1-n_\uparrow)^{L-N_\uparrow-1}P(L,N_\downarrow)}{\tilde{D}!n_\uparrow^{N_\uparrow}(1-n_\uparrow)^{L-N_\uparrow}P(L,N_\uparrow)}.$$

Then

$$\frac{\langle\psi|c_{i\uparrow}^\dagger c_{j\uparrow}|\psi\rangle}{\langle\psi|\psi\rangle} = \frac{(N_\uparrow-\tilde{D})(L-N_\uparrow-N_\downarrow+\tilde{D})}{L(L+1)}\frac{1}{n_\uparrow(1-n_\uparrow)}$$

or still

$$\frac{\langle\psi|c_{i\uparrow}^{\dagger}c_{j\uparrow}|\psi\rangle_1}{\langle\psi|\psi\rangle} = \frac{(n_\uparrow - \tilde{d})(1 - n_\uparrow - n_\downarrow + \tilde{d})}{n_\uparrow(1 - n_\uparrow)}.$$

Second Gutzwiller's Hopping

$$\downarrow\uparrow\ j\ \bullet - - - - - \bullet\ i\ \downarrow \Longrightarrow \downarrow\ j\ \bullet - - - - - - \bullet\ i\ \uparrow\downarrow$$

This is calculated as

$$\langle\psi|c_{i\uparrow}^{\dagger}c_{j\uparrow}|\psi\rangle = \sum_D \varrho^{2D+2} N_D(L-2, N_\uparrow - 1, N_\downarrow - 2) P(L-2, N_\uparrow - 1) P(L, N_\downarrow)$$

and yields

$$\frac{\langle\psi|c_{i\uparrow}^{\dagger}c_{j\uparrow}|\psi\rangle_2}{\langle\psi|\psi\rangle} = \varrho^2 \frac{(n_\uparrow - \tilde{d})(n_\downarrow - \tilde{d})^2}{n_\uparrow(1 - n_\uparrow)(1 - n_\uparrow - n_\downarrow + \tilde{d})}.$$

Third and Fourth Gutzwiller's Hoppings

$$\downarrow\uparrow\ j\ \bullet - - - - - \bullet\ i\ \Longrightarrow \downarrow\ j\ \bullet - - - - - - \bullet\ i\ \uparrow$$
$$\uparrow\ j\ \bullet - - - - - - \bullet\ i\ \downarrow \Longrightarrow\ j\ \bullet - - - - - - \bullet\ i\ \uparrow\downarrow$$

Now, in both cases, we have a doubly occupied site either before or after the jump. Both these contributions are calculated as

$$\langle\psi|c_{i\uparrow}^{\dagger}c_{j\uparrow}|\psi\rangle = \sum_D \varrho^{2D+1} N_D(L-2, N_\uparrow - 1, N_\downarrow - 1) P(L-2, N_\uparrow - 1) P(L, N_\downarrow)$$

to yield,

$$\frac{\langle\psi|c_{i\uparrow}^{\dagger}c_{j\uparrow}|\psi\rangle_{3,4}}{\langle\psi|\psi\rangle} = 2\varrho \frac{(n_\uparrow - \tilde{d})(n_\downarrow - \tilde{d})}{n_\uparrow(1 - n_\uparrow)}.$$

Note that in the calculations above, $P(L, N_\downarrow)$ is fixed, since only the spin $\uparrow$ hops. Also note that we have an additional factor ϱ for each doubly occupied site. If we now define q_σ as the sum of all hopping processes for a given spin σ, we get

$$q_\sigma = \frac{\langle\psi|c_{i\sigma}^{\dagger}c_{j\sigma}|\psi\rangle_{1+2+3+4}}{\langle\psi|\psi\rangle}$$

$$= \frac{n_\sigma - \tilde{d}}{n_\sigma(1 - n_\sigma)}\left[(1 - n_\sigma - n_{-\sigma} + \tilde{d}) + \varrho^2\frac{(n_{-\sigma} - \tilde{d})^2}{(1 - n_\sigma - n_{-\sigma} + \tilde{d})} + 2\varrho(n_{-\sigma} - \tilde{d})\right].$$

Using Eq. 10.1, this can be rewritten as

$$q_\sigma = \frac{\left[\sqrt{(n_\sigma - \tilde{d})(1 - n_\sigma - n_{-\sigma} + \tilde{d})} + \sqrt{(n_{-\sigma} - \tilde{d})\tilde{d}}\right]^2}{n_\sigma(1 - n_\sigma)}. \tag{10.2}$$

The factor q_σ essentially reduces the electronic hopping due to the restrictions imposed by the statistics and double occupancy.

Relevant Limits

1. Paramagnetic Metal ($n_\sigma = n_{-\sigma} = 1/2$)

$$q_\sigma = 8d(1 - 2d) \tag{10.3}$$

As in the equation above, from now on we use the notation d for $\tilde{d}$.

2. Strong Correlation ($U \rightarrow \infty, d \rightarrow 0$ and $n = n_\downarrow + n_\uparrow \leq 1$)

$$q_\sigma = \frac{1 - n}{1 - n_{-\sigma}} \tag{10.4}$$

This gives the behaviour of the hopping renormalisation factor close to the density-driven metal–insulator transition.

3. Polarised Fermi Liquid ($n_\sigma \simeq 1, n_{-\sigma} = d = 0$)
   In this case $q_\sigma \rightarrow 1$.

Properties of the Solution

The ground state energy per site can be written as

$$\frac{E_g}{L} = q_\uparrow(d, n_\uparrow, n_\downarrow)\epsilon_\uparrow + q_\downarrow(d, n_\uparrow, n_\downarrow)\epsilon_\downarrow + Ud \tag{10.5}$$

with $q_\sigma \geq 1$, the equality holding in the non-interacting case. The bare kinetic energies are given by

$$\epsilon_\sigma = \frac{1}{L}\langle\psi_0|\sum_{ij} t_{ij}c_{i\sigma}^\dagger c_{j\sigma}|\psi_0\rangle \tag{10.6}$$

$$= \sum_{|k|<k_f^\sigma} \epsilon_\sigma(k) < 0. \tag{10.7}$$

Ground State Energy

We consider initially a half-filled band, with $n = 1, n_\uparrow = n_\downarrow$ and $\epsilon_\uparrow = \epsilon_\uparrow = \epsilon_0/2$. In this case, the *hopping renormalisation factor q_σ* is independent of σ and given by

$$q = 8d(1 - 2d). \tag{10.8}$$

The ground state energy per site is

$$\frac{E_g}{L} = 8d\epsilon_0(1 - 2d) + Ud. \tag{10.9}$$

Minimising this energy with respect to the *double occupancy factor d*, we get

$$\frac{\partial(E_g/L)}{\partial d} = 8\epsilon_0 - 32d\epsilon_0 + U = 0$$

which gives

$$d = \frac{1}{4}\left(1 - \frac{U}{U_c}\right) \tag{10.10}$$

for $U \le U_c$ and $d = 0$ for $U > U_c$. The critical value of the Coulomb repulsion is, $U_c = 8|\epsilon_0|$. Substituting this in Eq. 10.8 for the hopping renormalisation factor, we get

$$q = 1 - \left(\frac{U}{U_c}\right)^2. \tag{10.11}$$

Consequently the effective hopping vanishes at the critical value U_c of the Coulomb repulsion, signalling a correlation-induced metal–insulator transition at U_c. Note that the double occupancy factor d in Eq. 10.10 plays the role of an order parameter, being non-zero in the metallic phase, i.e. for $U < U_c$ and zero in the insulating state ($U > U_c$).

Substituting the expression for the double occupancy factor, Eq. 10.10 in Eq. 10.9 for the ground state energy density, we find

$$\frac{E_g}{L} = -|\epsilon_0|\left(1 - \frac{U}{U_c}\right)^2. \tag{10.12}$$

From this equation and our previous definition of the critical exponent α associated with the singular part of the free or ground state energy density, we can identify the critical exponent $\alpha = 0$ within Gutzwiller's variational solution. The control parameter in this case is, $g = U - U_c$, such that we can write $f_s \propto |g|^{2-\alpha}$ with $\alpha = 0$.

Calculation of Thermodynamic Quantities

The general strategy to calculate the thermodynamic quantities is: given U, $n_\uparrow$, $n_\downarrow$, minimise the ground state energy, Eq. 10.5, with respect to d to obtain $d(n_\uparrow, n_\downarrow, U)$, $q_\sigma(n_\uparrow, n_\downarrow, U)$ and $E(n_\uparrow, n_\downarrow, U)$. The minimisation equation is

$$\sum_\sigma \epsilon_\sigma \frac{\partial q_\sigma}{\partial d} + U = 0.$$

From the equation for the energy $E(n_\uparrow, n_\downarrow, U)$, we then obtain the physical quantities of interest, namely, the chemical potential μ, the susceptibility χ and the compressibility κ. They are given by

$$\mu = \frac{\partial(E/L)}{\partial n} \tag{10.13}$$

$$\frac{1}{\chi} = \frac{\partial^2(E/L)}{\partial m^2} \tag{10.14}$$

$$\frac{1}{n^2\kappa} = \frac{\partial^2(E/L)}{\partial n^2} \tag{10.15}$$

where m is the magnetisation.

In the calculations to be carried out below we introduce a band model to calculate the energies ϵ_σ in Eq. 10.5. For simplicity we consider a constant density of states $\rho(\omega) = 1/2W = \rho_0$ from $-W$ to W and $\rho(\omega) = 0$ otherwise. Note, however, that the results we obtain for the critical exponents characterising the behaviour of physical quantities close to the metal–insulator transition are independent of a particular band-shape (see Lavagna, 1990).

Susceptibility

Let us consider a small external magnetic field which gives rise to a small magnetisation m in the system. The number of electrons of spin σ can be written as

$$n_\uparrow = \frac{1+m}{2}$$

$$n_\downarrow = \frac{1-m}{2}.$$

For $m \ll 1$ the chemical potentials for the up and down spin bands are shifted and given by

$$\mu(n_\uparrow) = \frac{m}{(2\rho_0/L)}$$

$$\mu(n_\downarrow) = \frac{-m}{2\rho_0/L}$$

such that

$$\epsilon(m) = \epsilon(n_\uparrow) + \epsilon(n_\downarrow) = \epsilon_0(1 + m^2).$$

From Eq. 10.2 for q_σ, we find

$$q_\uparrow = q_\downarrow = q = 4d\frac{1 - 2d + \sqrt{(1-2d)^2 - m^2}}{1 - m^2}$$

such that

$$\frac{E(m)}{L} = \frac{\epsilon(m)}{1 - m^2} 4d[1 - 2d + \sqrt{(1 - 2d)^2 - m^2}] + Ud. \qquad (10.16)$$

We have to consider two cases:

- $U > U_c$ In this case, $d = 0$, $E(m) = $ constant and this yields an infinite susceptibility in the insulating phase. As pointed out before, this is not due to interactions but a consequence of the appearance of local moments for which the susceptibility diverges at zero temperature.
- $U < U_c$ In the metallic state, $U < U_c$, the energy E depends directly on the magnetisation m, as shown in the equation above and through $d(m)$, which is obtained through the minimisation condition, $(\partial E(m, d)/\partial d) = 0$. Using this condition, we obtain that $(\partial E(m, d)/\partial m) = (\partial E(m, d)/\partial m)_d$. Furthermore, using $d(m) = d(m = 0) + am^2$, such that $(\partial d/\partial m)_{m=0} = 0$, it turns out that it is sufficient to differentiate Eq. 10.16 with respect to m keeping d fixed. We finally get for the susceptibility

$$\frac{1}{\chi} = \frac{\partial^2(E/L)}{\partial m^2}$$

$$= 2q\epsilon_0 \frac{-1}{(1 + u)^2}$$

where $q = 8d(1 - 2d)$ and $u = U/U_c$. Note, then, that the divergence of χ is related to the vanishing of the renormalisation hopping factor q at the metal–insulator transition. In this sense, there is no special critical exponent related to the divergence of χ as in a magnetic transition. The present divergence is a consequence of the formation of non-interacting local moments at U_c and $T = 0$.

Nearly Half-Filled Band and Arbitrary U

Let us write

$$n_\uparrow = n_\downarrow = \frac{1 - \delta}{2}$$

such that, $n = n_\uparrow + n_\downarrow = 1 - \delta$ where $\delta \ll 1$ measures the deviation from half-filling. In this case we always have a metallic ground state independently of U. The hopping renormalisation factor can be written as

$$q_\uparrow = q_\downarrow = q = \frac{2(1 - \delta - 2d)}{1 - \delta^2} \left[\sqrt{d} + \sqrt{\delta + d}\right]^2.$$

We also find, for the bare kinetic energy,

$$\epsilon_\uparrow = \epsilon_\downarrow = \frac{1}{2}\epsilon_0(1 - \delta^2).$$

Now we follow the procedure of minimising E with respect to d to obtain $d(\delta, U)$, $q(\delta, U)$ and finally $E(\delta, U)$. It turns out to be useful to make the following change of variable:

$$x = \sqrt{d} + \sqrt{d + \delta} \tag{10.17}$$

such that

$$d = \left[\frac{x^2 - \delta}{2x}\right]^2$$

and

$$q = \frac{2x^2 - x^4 - \delta^2}{1 - \delta^2}$$

and now x substitutes d as the variational parameter. The energy per site can be written in terms of x as

$$\frac{E}{L} = \epsilon_0\left[2x^2 - x^4 - \delta^2\right] + U\frac{(x^2 - \delta)^2}{4x^2}. \tag{10.18}$$

The minimisation condition, $(\partial(E/L)/\partial x) = 0$, yields

$$\frac{x^4(1 - x^2)}{x^4 - \delta^2} = \frac{U}{U_c} = u \tag{10.19}$$

with $U_c = 8|\epsilon_0|$.

Chemical Potential and Compressibility for $U \leq U_c$

We start from Eq. 10.18 and, noticing from Eq. 10.19 that $= x \approx 1 - u + O(\delta^2)$, we obtain the energy to $O(\delta^2)$, as required for the calculation of the compressibility:

$$\frac{E}{L} = \epsilon_0(1 - u)^2 - \frac{U}{2}\delta - \epsilon_0\delta^2\frac{1 - u^2}{(1 - u)^2}.$$

The chemical potential

$$\mu = \frac{dE}{dN} = -\frac{d(E/L)}{d\delta} = \frac{U}{2}$$

which is the exact result for the chemical potential in the half-filled band case. The compressibility κ is given by

$$\frac{v}{n^2\kappa} = \frac{d\mu}{dn} = \frac{\partial^2(E/L)}{\partial\delta^2}$$

where v is the volume. This is finally obtained as

$$\frac{\kappa_0}{\kappa} = \frac{1 - u^2}{(1 - u)^2} \propto \frac{1}{1 - u}$$

where $\kappa_0 = 1/2|\epsilon_0|$ is the compressibility of the non-interacting system. We find then that in Gutzwiller's approach the compressibility vanishes at the correlation-induced MI transition, with the distance to the critical point, i.e. $\kappa \propto |g|$ where $g = U - U_c$.

Chemical Potential for $U \geq U_c$

For $u > 1, d \propto \delta$ (see below) and from Eq. 10.17, we get, $x \propto \sqrt{\delta}$. Using Eq. 10.19 we obtain

$$x^2 = \frac{\delta}{\sqrt{1 - \frac{1}{u}}}.$$

We also find

$$q = 2x^2 = \frac{2\delta}{\sqrt{1 - \frac{1}{u}}} \tag{10.20}$$

and

$$d = \delta \frac{[1 - \sqrt{1 - \frac{1}{u}}]^2}{4\sqrt{1 - \frac{1}{u}}}.$$

Since q and d are proportional to δ, it is the presence of the holes that makes the system conductor. We need now the energy to $O(\delta)$ to calculate the chemical potential; this is given by Eq. 10.18,

$$\frac{E}{L} = 2\epsilon_0 x^2 + U \frac{(x^2 - \delta)^2}{4x^2} \tag{10.21}$$

which can be rewritten as

$$\frac{E}{L} = -\frac{U\delta}{2} \left(1 - \sqrt{1 - \frac{1}{u}} \right) \tag{10.22}$$

since, $\mu = -d(E/L)/d\delta$, we find

$$\frac{\mu}{U} = \frac{1}{2} \pm \frac{1}{2}\sqrt{1 - \frac{1}{u}},$$

where the $+$ sign has been introduced due to electron–hole symmetry. This corresponds to the situation where electrons are removed (added) to the half-filled

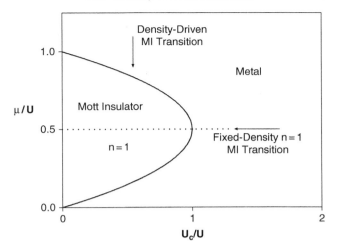

Figure 10.2 Phase diagram of the Hubbard model within the Gutzwiller approximation. Both the $n = 1$ commensurate fixed-density and the density-driven metal–insulator transitions are indicated. In the former, the chemical potential is kept fixed at $\mu = U/2$, which corresponds to $n = 1$, while the Coulomb repulsion increases. In the latter, U is fixed while the density of electrons is changed (Fig. 1 adapted with permission from Continentino, M. A. (1991), Phys. Rev. B43, 6292).

band. These equations give the boundaries of the Mott insulating phase shown in Fig. 10.1. Consequently at $n = 1$ there is a gap either to remove or add an electron to the system. The gap is given by

$$\Delta = \sqrt{1 - \frac{1}{u}} = \sqrt{\frac{U - U_c}{U_c}}. \tag{10.23}$$

Thus the gap vanishes at the critical value U_c with a *gap exponent $vz = 1/2$*. So we have identified another important exponent within the Gutzwiller approximation.

10.3 Density-Driven Transition

Now to $O(\delta^2)$, the ground state energy per site for $U > U_c$ is given by

$$\frac{E}{L} \approx -\frac{U}{2}\left(1 - \sqrt{1 - \frac{1}{u}}\right)\delta - \epsilon_0\left(1 + \frac{1}{1 - \frac{1}{u}}\right)\delta^2.$$

The free energy density is $f = E/L - \mu n$. Minimising the free energy with respect to δ we find the chemical potential to $O(\delta)$ and the relation

$$\mu - \mu_c(U) = \frac{1}{2\epsilon_0}\left[1 + \frac{1}{1 - \frac{1}{u}}\right]\delta, \tag{10.24}$$

where

$$\mu_c(U) = \frac{U}{2} \pm \frac{U}{2}\sqrt{1 - \frac{1}{u}} \qquad (10.25)$$

give the equations for the critical boundaries of the Mott insulating phase for $U > U_c$ (see Fig. 10.1). Substituting Eq. 10.24 above in the free energy, we get

$$f \propto A(U)\,[\mu - \mu_c(U)]^2 . \qquad (10.26)$$

From this equation we identify the exponent $\tilde{\alpha} = 0$ for the density-driven MI transition. The amplitude $A(U)$ is given by

$$\frac{1}{A(U)} = 4\epsilon_0[1 + \frac{1}{1 - \frac{1}{u}}].$$

The compressibility, $\kappa = -(\partial^2 f/\partial\mu^2)$ is given by

$$\frac{1}{2\kappa} = |\epsilon_0|\left[1 + \frac{1}{1 - \frac{1}{u}}\right]. \qquad (10.27)$$

In this case the compressibility attains a finite value at the metal–insulator transition as it is approached from the metallic side. On the other side of the transition, the Mott insulating phase satisfies a commensurability condition with exactly one electron per site. This Mott phase exists for a finite range of the chemical potential as can be seen in Fig. 10.1. Within this range, for a fixed $U > U_c$, the chemical potential varies but n is fixed at the commensurate value $n = 1$. Consequently in the Mott phase, the compressibility $\kappa = \partial n/\partial\mu$ vanishes and this phase is *incompressible*. Then, along a fixed $U > U_c$ trajectory, the compressibility has a jump at the density-driven MI transition associated with the exponent $\tilde{\alpha} = 0$.

10.4 Scaling Analysis

Correlation-Induced or Fixed-Density Mott Transition

Let us consider first the correlation-induced MI transition that occurs as a function of the Coulomb term U for fixed $n = 1$. We have identified for this transition the critical exponents, $\alpha = 0$ (Eq. 10.12) and the gap exponent, $\nu z = 1/2$ (Eq. 10.23). In order to calculate the correlation length exponent independently, we need an additional assumption relating the hopping renormalisation parameter q to the conductivity mass m_c, i.e. we take $m_c \propto 1/q$ or $D_c \propto q$ where D_c is the charge stiffness. Since we have obtained $q \propto |g| = |U - U_c|$ (Eq. 10.11) and according to Eq. 8.11, the charge stiffness scales as $D_c \propto |g|^{2-\alpha-2\nu}$, using $\alpha = 0$ we find $\nu = 1/2$ and, consequently, $z = 1$. These values for the critical exponents α, ν and z are typical mean field values. Also it is clear from the calculations above that dimensionality plays no role in the Gutzwiller approach. This suggests that this approach represents the mean field solution of the correlation-induced Mott transition problem. When the Gutzwiller exponents are substituted in the quantum hyperscaling relation, $2 - \alpha = v(d + z)$, we find that it is satisfied for $d = d_c = 3$. *Then, the scaling argument yields that $d_c = 3$ is the upper critical dimension for the correlation-induced Mott insulator transition above which the Gutzwiller approximation gives the correct description of this phenomenon.* The scaling form of the free energy close to the quantum critical point at $U = U_c$, $\mu = U/2$ is given by

$$ f \propto |g|^{2-\alpha} F \left[\frac{\delta \mu}{|g|^{vz}} \right] \tag{10.28} $$

with $\delta \mu = \mu - U/2$, $\alpha = 0$ and $vz = 1/2$. The compressibility, $\kappa \propto \partial^2 f / \partial \mu^2$ and we get $\kappa \propto |g|$ at the fixed-density MI transition as we have found previously.

Density-Driven Mott Transition

The phase transition described by Eq. 10.26 is a density-driven metal–insulator transition since we approach the boundary of the Mott phase, at $\mu_c(U)$, by varying the chemical potential or particle density *at a constant value U of the Coulomb repulsion*. Again, in this case we find that the critical exponent $\tilde{\alpha} = 0$. Using that the charge stiffness D_c is directly proportional to the hopping renormalisation parameter q, that the former scales as $D_c \propto |g|^{2-\tilde{\alpha}-2\tilde{v}}$ and Eqs. 10.20 and 10.24 which lead to $q \propto |\mu - \mu_c(U)|$, we find $\tilde{v} = 1/2$. Furthermore, since $\tilde{v}\tilde{z} = 1$ for the density-driven transition (Continentino, 1994; Imada *et al.*, 1998), we find $\tilde{z} = 2$. Note that for this density-driven transition the control parameter is $|g| = |\mu - \mu_c(U)|$. Besides, the compressibility given by Eq. 10.27 is finite at the MI transition, and vanishes in the Mott phase. This is different from the behaviour at the fixed-density MI transition, where the compressibility vanishes at the transition when approaching it from the metallic side. The finite compressibility at the transition and the scaling expression, $\kappa \propto |g|^{\tilde{v}(d-\tilde{z})}$ (Imada *et al.*, 1998), implies the additional relation $\tilde{z} = d$ (Fisher *et al.*, 1989). In this case, $d = 2$, the dimension for which this equality is satisfied, has the meaning of an upper critical dimension for this density-driven transition in the presence of interactions. Then differently from the results of the previous chapter, our scaling analysis of Gutzwiller's solution suggests the existence of an upper critical dimension, $\tilde{d}_c = 2$, for density-driven transitions in the interacting problem. This is further supported by the mean field-like values of the exponents $\tilde{\alpha}$ and $\tilde{v}$ which imply that the Gutzwiller solution is the mean field solution of the Mott transition problem. Notice that $\tilde{d}_c = 2$ is consistent with the results for the density-driven transition in the $d = 1$ Hubbard model since $d < \tilde{d}_c$. We point out, however, that there is numerical evidence that the Mott transition in $d = 2$ is in a different universality class of that obtained above (see Imada *et al.*, 1998).

Critical Trajectory

It is not difficult, using the arguments developed in Chapter 1 and Eq. 10.28, to obtain the physical behaviour along the special trajectory, $U = U_c$, $\delta\mu \to 0$. This is a special density-driven transition since the Coulomb interaction U is fixed. Extended scaling applies since, $\psi = \nu z = 1/2$, the shift exponent ψ being defined by the shape of the phase boundary between the metal and the Mott insulator close to U_c (see Eq. 10.25). The scaling form of the free energy, $f \propto |g|^{2-\alpha} F[\delta\mu/|g|^{\nu z}]$ and the expressions, $\kappa = (\partial^2 f/\partial\mu^2)$, $\delta \propto (\partial f/\partial\mu)$, together with extended scaling leads to, $\kappa \propto \delta\mu^{(2-\alpha-2\nu z)/\nu z} = \delta\mu^2$ and $\delta \propto \delta\mu^{(2-\alpha-\nu z)/\nu z} = \delta\mu^3$. Notice that the critical exponents are those associated with the fixed-density transition. This finally yields, $\kappa \propto \delta^{2/3}$ at $U = U_c$, where δ, we recall, is the deviation from half-filling and $\delta\mu = (\mu - U/2)$.

The extension to finite temperatures of the Gutzwiller approach is not a trivial task and must be supplemented by additional assumptions. Let us consider the calculation of the specific heat of the highly correlated metal close to the fixed-density metal insulator transition at U_c. According to the scaling theory developed in previous chapters, the specific heat below the coherence line, $T_{coh} \propto |g|^{\nu z}$, is linear in temperature and given by

$$C/T \propto |g|^{2-\alpha-2\nu z}$$

where $|g| \propto |U - U_c|$. Since, as we have shown, the critical exponents associated with the quantum critical point at $\mu = U/2$, $U = U_c$, which governs the zero temperature fixed-density metal–insulator transition are $\alpha = 0$ and $\nu z = 1/2$, we find that the scaling contribution to the thermal mass $m_T \propto C/T \propto |g|$ actually vanishes at the transition. On the other hand, if we assume that the thermal mass is proportional to the conductivity mass m_c, which scales as, $m_c \propto 1/q \propto |U - U_c|^{-1}$, we find a diverging mass at the transition. As for the density driven transition, the expression above with the *tilde* exponents, $\tilde\alpha = 0$ and $\tilde\nu\tilde z = 1$, suggest a finite thermal mass, or at most a logarithmic divergence of this mass, for all $d \geq 2$ at the density driven Mott transition.

10.5 Conclusions

In this chapter we have presented a scaling analysis of the Gutzwiller solution for the Mott transition in the Hubbard model. From this analysis we were able to identify the critical exponents associated with this solution, for both the density-driven and correlation-induced or fixed-density transition. The two transitions turn out to be in different universality classes. The Gutzwiller solution is clearly a mean field approximation and from the knowledge of its critical exponents and a scaling argument, we were able to identify the upper critical dimensions, $d_c = 3$ and $\tilde{d}_c = 2$ for the fixed-density and density-driven transitions, respectively. For the latter we have found that interactions can modify the critical behaviour with respect to that of the non-interacting case, contrary to what we would have expected from the exact results in the infinite range model and the $d = 1$ Hubbard model of the previous chapter.

Recently there has been renewed interest in the problem of the Mott transition in part due to progress in the solution of this problem at $d = \infty$. The picture which arises from this solution is quite different (Georges, 1996). In this case the $T = 0$ Mott transition turns out to be first order, which in principle prevents a scaling analysis.

11

The Non-Linear Sigma Model

11.1 Introduction

The non-linear sigma model plays an important role in several problems in condensed matter physics, such as the localisation transition, localised magnetism and superconductivity (Sachdev, 1999; Auerbach, 1994; Fradkin, 1991; Chakravarty, *et al.*, 1988). For such relevance, and also for fixing and further clarifying concepts dealt with in this book, we study it now. For our purposes, we will consider this model essentially as a continuum mapping of the quantum antiferromagnet containing the basic features of rotational invariance and linear spin-wave spectrum. The treatment we present is an ϵ-expansion close to the *lower critical dimension*, d_L, in contrast to what we have done in Chapter 3, where the expansion was near the upper critical dimension, $d_c = 4$.

The action of the non-linear sigma model is given by

$$S = \frac{1}{2\varrho} \int d^d x \, (\nabla \vec{m})^2 .$$ (11.1)

The non-linear character is given by the constraint

$$|\vec{m}|^2 = 1,$$ (11.2)

which mixes the different spin components. Introducing a dimensionless coupling constant v by the equation

$$v = \varrho a^{2-d},$$

we rewrite the action 11.1 as

$$S = \frac{1}{2va^{d-2}} \int d^d x \, (\nabla \vec{m})^2 .$$ (11.3)

The quantity a^{-1} represents an ultraviolet cut-off since a is the lattice spacing.

The quantum non-linear sigma model is just the classical model above in $d + 1$ dimensions. The reason is that the dynamic exponent takes the value $z = 1$ by the definition of the dynamics of this model, as discussed later in this chapter. Then, we are dealing here with a Lorentz invariant action where *time* just enters as an additional dimension.

11.2 Transverse Fluctuations

We assume the system has short-range order such that the spins are mostly parallel to a given direction. For the sake of clarity we consider here the specific case of three dimensions, $d = 3$, and take the direction $\hat{z}$ as the preferential direction of the spins. The constraint, Eq. 11.2, can be written as

$$m_z^2 + S^2 = 1 \tag{11.4}$$

such that

$$S = \sqrt{1 - m_z^2}$$

with

$$m_x = S \cos \theta = \sqrt{1 - m_z^2} \cos \theta$$

and

$$m_y = S \sin \theta = \sqrt{1 - m_z^2} \sin \theta.$$

Of course, $m_x^2 + m_y^2 + m_z^2 = 1$. Let us consider the Lagrangian density:

$$L = \frac{1}{2va^{d-2}} (\nabla \vec{m})^2 = \frac{1}{2va^{d-2}} \left[(\nabla m_z)^2 + \left(\nabla m_y \right)^2 + \left(\nabla m_x \right)^2 \right].$$

It is easy to check, with the help of Fig. 11.1, that

$$(\nabla m_x)^2 = \frac{(m_z \nabla m_z)^2}{1 - m_z^2} \cos^2 \theta + (1 - m_z^2)(\nabla \theta)^2 \sin^2 \theta$$

and

$$\left(\nabla m_y \right)^2 = \frac{(m_z \nabla m_z)^2}{1 - m_z^2} \sin^2 \theta + (1 - m_z^2)(\nabla \theta)^2 \cos^2 \theta.$$

Adding these contributions we get

$$L = \frac{1}{2va^{d-2}} \left[(\nabla m_z)^2 + (1 - m_z^2)(\nabla \theta)^2 + \frac{(m_z \nabla m_z)^2}{1 - m_z^2} \right].$$

Introducing a new field ϕ, such that, $m_z = \sqrt{va^{d-2}}\phi$ the Lagrangian density can be written as

$$L = \frac{1}{2}(\nabla \phi)^2 + \frac{1 - va^{d-2}\phi^2}{2va^{d-2}}(\nabla \theta)^2 + \frac{va^{d-2}}{2(1 - va^{d-2}\phi^2)}(\phi \nabla \phi)^2. \tag{11.5}$$

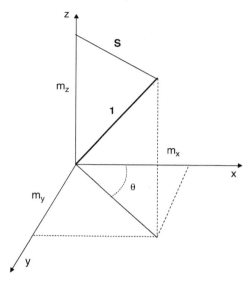

Figure 11.1 Components of the local magnetisation for the 3d classical non-linear sigma model.

For small $v << 1$, we can approximate the above expression by

$$L = \frac{1}{2}(\nabla\phi)^2 + \frac{1}{2va^{d-2}}(\nabla\theta)^2 - \frac{1}{2}\phi^2(\nabla\theta)^2 + \frac{1}{2}va^{d-2}\phi^2(\phi\nabla\phi)^2$$
$$+ \frac{1}{2}v^2a^{2(d-2)}\phi^2(\phi\nabla\phi)^2 + 0(v^3).$$

We will consider here the action

$$S = \int d^d x \left[\frac{1}{2}(\nabla\phi)^2 + \frac{1}{2va^{d-2}}(\nabla\theta)^2 - \frac{1}{2}\phi^2(\nabla\theta)^2 + O(v)\right] \tag{11.6}$$

and the following functional integral, where the short wavelength modes are removed

$$\int_{b\Lambda < |\vec{k}| < \Lambda} D\phi(k)e^{-S[\phi(k),\theta]}.$$

We assume that the variation of θ is sufficiently slow that it has no components in the interval $b\Lambda < |\vec{k}| < \Lambda$. In this case, the functional integral becomes

$$\int_{b\Lambda < |\vec{k}| < \Lambda} D\phi(k)e^{-\frac{1}{2}\int \frac{d^d k}{(2\pi)^d}[|\vec{k}|^2 + (\nabla\theta)^2]|\phi(k)|^2} \approx \prod_{b\Lambda < |\vec{k}| < \Lambda} \left[\frac{2\pi}{|\vec{k}|^2 + (\nabla\theta)^2}\right]^{1/2}.$$

This may still be written as

$$exp\left(\int_{b\Lambda < |\vec{k}| < \Lambda} \frac{d^d k}{(2\pi)^d} Ln\left[\frac{2\pi}{|\vec{k}|^2 + (\nabla\theta)^2}\right]^{1/2}\right) \approx$$

$$exp\left(\int_{b\Lambda<|\vec{k}|<\Lambda}\frac{d^dk}{(2\pi)^d}Ln\left[\frac{2\pi}{|\vec{k}|^2}\right]+\frac{1}{2}\int_{b\Lambda<|\vec{k}|<\Lambda}\frac{d^dk}{(2\pi)^d}Ln\left[\frac{1}{1-\frac{(\nabla\theta)^2}{|\vec{k}|^2}}\right]\right)$$

or still

$$exp\left(\int_{b\Lambda<|\vec{k}|<\Lambda}\frac{d^dk}{(2\pi)^d}Ln\left[\frac{2\pi}{|\vec{k}|^2}\right]+\frac{1}{2}(\nabla\theta)^2\int_{b\Lambda<|\vec{k}|<\Lambda}\frac{d^dk}{(2\pi)^d}\frac{1}{|\vec{k}|^2}\right).$$

The first term represents just a shift in the energy with respect to the original action, Eq. 11.6. The term in $(\nabla\theta)^2$ has a coefficient different from that of the action, Eq. 11.6. Then, the effect of integrating the short wavelength modes is to renormalise the stiffness of the θ modes, essentially the coupling constant v. Performing the k-integration in the equation above and comparing the coefficient of the $(\nabla\theta)^2$ term with that of Eq. 11.6, we define the renormalised stiffness through the equation

$$\frac{1}{v_{n+1}a'^{(d-2)}}=\frac{1}{v_na^{d-2}}-\frac{S_d}{(2\pi)^d}\frac{1-b^{d-2}}{d-2}\Lambda^{d-2}$$

where $a'=a/b$ and the last term comes from the integral of the second term in the exponential above. We get

$$\frac{1}{v_{n+1}}=\frac{1}{v_n}\left(\frac{a'}{a}\right)^{d-2}\left[1-v_n\frac{S_d}{(2\pi)^d}\frac{1-b^{d-2}}{d-2}\right]$$

where we used that $\Lambda=1/a$. For v small the recursion relation can be written as

$$v_{n+1}=b^{d-2}v_n+\frac{S_d}{(2\pi)^d}\frac{b^{d-2}-b^{2(d-2)}}{d-2}v_n^2.$$

In order to find the continuous version of the renormalisation group equation, we write $b^{2-d}=\exp[(2-d)\ln b]\approx 1+(2-d)\ln b$, where in the last step the limit $b\to 1$ was taken. We get

$$\frac{\partial v}{\partial\ln b}=(2-d)v+\frac{S_d}{(2\pi)^d}v^2\equiv\beta(v).$$

This last identity defines the *beta-function*. The fixed points, v_i^*, of the RG equations correspond to the zeros of the beta function, i.e. $\beta(v_i^*)=0$. This equation leads to two fixed points, namely, $v_0^*=0$ and $v_1^*=(d-2)(2\pi)^d/S_d$. For $d\approx 2$, both fixed points are located at small values of v justifying the expansions we have made.

The stability of the fixed points is studied expanding the RG equations in their neighbourhood:

$$\frac{\partial v}{\partial\ln b}\approx\beta(v^*)+\left(\frac{\partial\beta}{\partial v}\right)_{v=v^*}(v-v^*)$$

where

$$\frac{\partial \beta}{\partial v} = (2 - d) + \frac{2S_d}{(2\pi)^d} v.$$

Then

$$\left(\frac{\partial \beta}{\partial v}\right)_{v=v_0^*=0} = 2 - d$$

and since $\beta(v^*) = 0$, this is a stable fixed point for $d > 2$ and unstable for $d < 2$. Also,

$$\left(\frac{\partial \beta}{dv}\right)_{v=v_1^*} = d - 2,$$

which in turn is an unstable fixed point, i.e. the flow of the renormalisation group equation is away from this fixed point for $d > 2$ and it is stable, or attractive, for $d < 2$.

The case $d = 2$ is *marginal*. The marginal behaviour is associated with the collapse of the two fixed points $v_0^* = v_1^* = 0$. In this case, the first term of the expansion of the beta function is non-linear in the coupling v. Notice that there is an exchange of stability at $d = 2$ and the fixed point at $v^* = 0$ is now unstable.

We now integrate the RG equations close to the unstable fixed point to obtain the behaviour of the correlation length. In order to be consistent with our assumption of v small, we require $d \approx 2$, such that, $d = 2 + \epsilon$ with ϵ small. For $\epsilon > 0$ the unstable fixed point is then at $v_1^* \approx 2\pi\epsilon$. In this case we have

$$\frac{\partial(v - v^*)}{\partial \ln b} = (d - 2)(v - v^*).$$

Integrating from $b = L_0$ to $b = L$, we get

$$\ln\left[\frac{v(L) - v^*}{v(L_0) - v^*}\right] = (d - 2)\ln\left(\frac{L}{L_0}\right).$$

For $v(L_0) = v_0$, close to the critical point, $L = \xi$ is given by

$$\xi \propto \left|\frac{1}{v_0 - v^*}\right|^{-\nu} \tag{11.7}$$

with $\nu = 1/(d - 2)$.

In the marginal case, $d = 2$, where $v_1^* = 0$, we have

$$\frac{\partial v}{\partial \ln b} = \frac{v^2}{2\pi}.$$

Integrating from $b = L_0$ to $b = L$, we obtain

$$v(L) = \frac{v_0}{1 + \frac{v_0}{2\pi}\ln\left(\frac{L_0}{L}\right)}.$$

For $L = L_0$, v_0 is small, but when L grows $v(L)$ also increases. This phenomenon, in which the coupling increases as the distance L increases, is known as *asymptotic freedom* (Fradkin, 1991). Note that at a characteristic length $L = \xi$ given by

$$1 + \frac{v_0}{2\pi} \ln\left(\frac{L_0}{\xi}\right) = 0$$

or

$$\xi = L_0 e^{\frac{2\pi}{v_0}} \tag{11.8}$$

the weak coupling expansion breaks down as v becomes very large.

For $d < 2$, the unstable fixed point is now at $v^* = 0$. The RG equation close to this fixed point can be easily integrated and we obtain, close to transition at $v_0 = 0$,

$$\xi \propto \left(\frac{1}{v_0}\right)^{\frac{1}{|\epsilon|}}. \tag{11.9}$$

The type of behaviour obtained above for the correlation length, namely power law dependence on the control parameter below and above the lower critical dimension d_c, Eqs. 11.9 and 11.7, respectively and exponential at d_c, Eq. 11.8, is a general feature of phase transitions. We recall the definition of lower critical dimension, d_L, as that above which the transition occurs at a finite value of the control parameter: in the present case, the coupling v.

11.3 The Quantum Non-Linear Sigma Model

We will consider now explicitly the quantum problem and an important application (Chakravarty, *et al.*, 1988). The action for the case of a three-component vector field is given by

$$\hbar^{-1}S = \frac{\rho_s^0}{2\hbar} \int d^d x \int_0^{\hbar/k_B T} d\tau \left[\frac{1}{c^2}|\partial_\tau \vec{m}|^2 + |\partial_x \vec{m}|^2 + |\partial_y \vec{m}|^2 + |\partial_z \vec{m}|^2\right]. \tag{11.10}$$

We have written this action in a form that shows explicitly the Lorentz invariance of the model with *time*, in fact ($c\tau$), acting just as an additional dimension. Notice that at finite temperatures the integral in the time direction is cut off at $\hbar/k_B T$. The quantity ρ_s^0 is a generalised stiffness which plays the role of the control parameter. We will be interested in the two-dimensional case, since this corresponds to the $d = 2$ Heisenberg antiferromagnet, which in turn is relevant in the study of high temperature superconductors. At $T = 0$ this model corresponds to the classical model studied in the previous session in $d+1$ dimensions. Then at $T = 0$ and $d = 2$ there is a quantum critical point for a finite value of the control parameter with a correlation length exponent $\nu = 1/(d - 1)$ as given by Eq. 11.7. Let us define g as

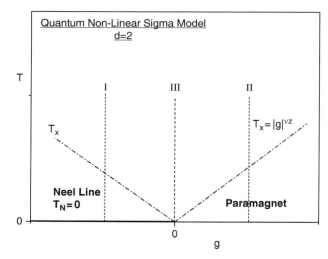

Figure 11.2 Phase diagram of the quantum non-linear sigma model (Chakravarty, Halperin and Nelson, 1988).

the $T = 0$ distance to the QCP. Since the stiffness scales as $\rho_s^0 \propto |g|^{\nu(d+z-2)}$, we get $\rho_s^0 \propto |g|$, proportional to the distance to the QCP. The phase diagram for the quantum non-linear sigma model at $d = 2$ is shown in Fig. 11.2. The crossover lines $T_{cross} \propto |g|^{\nu z} = |g|$ and since the theorem of Mermin–Wagner (Stanley, 1971) precludes magnetic order in $d = 2$ at finite temperatures, the ordered antiferromagnetic phase can only exist at $T = 0$ and the Néel line, $T_N = 0$, is flat as shown in Fig. 11.2.

Consider reducing the temperature along the paths shown in Fig. 11.2. For trajectory I, this corresponds to lowering the temperature in a system which has order only at $T = 0$ and consequently we expect an exponential temperature dependence for the correlation length, $\xi \propto \exp(|g|/T)$, as in the case of the previous section where the system is at the lower critical dimension (Eq. 11.8). In the disordered phase, decreasing the temperature along path II yields a finite correlation length $\xi \propto |g|^{-\nu}$ at $T = 0$. Furthermore, below the crossover line $T_{cross} \propto |g|^{\nu z} = |g|^{1/(d-1)}$, the physical properties are dominated by a gap as in the transverse Ising model and we expect exponentially small corrections to this value at finite $T < T_{cross}$.

Along the critical trajectory III, we obtain the behaviour of the correlation length with the *ansatz* for the scaling function used before, i.e. we start with

$$\xi = |g|^{-\nu} f\left[\frac{T}{|g|^{\nu z}}\right]$$

and require that $f[y] \propto y^x$ where x is such that the dependence on g cancels out

$$\xi \propto |g|^{-\nu}\left(\frac{T}{|g|^{\nu z}}\right)^x .$$

Since $z = 1$, the requirement that g cancels out implies that $x = -1$ and we obtain $\xi \propto 1/T$. This type of argument represents a *tour de force* of the scaling theory. Although it gives the appropriate description of the critical behaviour of the model, it does not imply we can prescind from detailed calculations, being useful as a general approach. Results of calculations of the correlation length in the quantum non-linear sigma model when compared to experimental data on La_2CuO_4, the parent compound of a class of high temperature superconductors, have shown that on cooling this system the trajectory is as in *I* (Chakravarty, *et al.*, 1988). This is indicated by the temperature dependence of the correlation length and establishes the nature of the ground state of a CuO_2 plane as being a long-range ordered two-dimensional antiferromagnet.

11.4 Some Notable β-Functions

As for the non-linear sigma model, many important features of a physical problem are synthesised in their β-functions. We discuss now some important cases where the knowledge of the β-functions yields to very general and universal results for the system.

Anderson Localisation

In this case the β-function describes the behaviour of the conductance g, a dimensionless quantity of a disordered non-interacting electronic system at the length scale L (Lee and Ramakrishnan, 1985). The renormalisation group equation for g is given by

$$\frac{d \ln g}{d \ln L} = \beta(g) \qquad (11.11)$$

where the β-function obtained perturbatively for weak disorder is given by

$$\beta(g) = (d - 2) - \frac{a}{g}, \qquad (11.12)$$

with $a = \pi^{-2}$ for an electron gas. This function is shown in Fig. 11.3 for one, two and three dimensions. For $d = 3$ the beta function has a zero at $g = a = g_c$, implying the existence of a fixed point of the flow equation. This is an unstable fixed point, which is physically associated with a mobility edge or a disorder-driven metal–insulator transition at a critical value g_c of the conductance. For $d = 1$ all states are localised, for arbitrarily weak disorder, and $d = 2$ represents the marginal case.

For $d = 2 + \epsilon$, the fixed point at $g = g_c = a/\epsilon$ is fully unstable. Expanding close to it, one obtains a correlation length, to be interpreted as a localisation length, which diverges as $\xi_L \propto |g - g_c|^{-\nu}$, with $\nu = 1/\epsilon$. In this case, for $3d$ the correlation length exponent turns out to be $\nu = 1$.

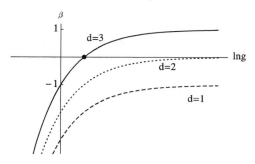

Figure 11.3 The β-function for the Anderson localisation problem in one, two and three dimensions (schematic). For $d = 3$, β has a non-trivial zero associated with a metal–insulator transition at a critical value of the conductance g_c, such that $\beta(g_c) = 0$.

At the marginal dimension $d = 2$, the β-function can be integrated to give a length-dependent conductance

$$g(L) = g_0 - \frac{e^2}{\hbar\pi^2} \ln(\frac{L}{l}), \tag{11.13}$$

where g_0 is the conductance at scale l.

Although the β-function yields a lot of information on the localisation transition, this is a quantum phase transition and as we have seen the dynamical aspects of this transition must be considered to yield a complete description of the critical behaviour. For the Anderson transition the dynamic exponent has been obtained through the calculation of the frequency dependent conductivity and turns out to be $z = d$ (Lee and Ramakrishnan, 1985; Fisher *et al.*, 1989, 1990). We point out that in spite of this equality, contrary to the case of magnetic transitions where it yields to logarithmic behaviour, for example, in thermodynamic quantities, in the localisation problem the logarithms appear only at the critical lower dimension, $d = 2$ as in Eq. 11.13 in spite that $z = d$ for any d.

Quantum Electrodynamics

In the case of quantum electrodynamics (QED), the coupling constant, i.e. the electric charge or as more commonly used the fine-structure *constant* α is a scale dependent quantity (see Huang, 2015). Expressing it in terms of momentum scale Λ, instead of a length scale, it turns out to obey a renormalisation group equation given by

$$\Lambda^2 \frac{\partial \alpha(\Lambda^2)}{\partial \Lambda^2} = \beta(\Lambda^2), \tag{11.14}$$

where $\beta = \beta_{QED}$ is a positive quantity given by

$$\beta_{QED} = \frac{\alpha^2(\Lambda^2)}{3\pi}. \tag{11.15}$$

This type of RG equation describes the flow of a marginally relevant interaction which iterates to strong coupling at large momenta or small distances. The only (unstable) fixed point corresponds to the trivial one at $\alpha = 0$. At least at this level of approximation there is no phase transition besides this trivial zero of β.

An interesting feature related to this β-function is the presence of a Landau *ghost* that led to many misconceptions in the development of quantum field theory. This can be materialised integrating the RG equation using the β-function given by Eq. 11.15. We get

$$\ln \frac{\Lambda^2}{\Lambda_0^2} = 3\pi \int_{\alpha(\Lambda_0^2)}^{\alpha(\Lambda^2)} \frac{d\alpha}{\alpha^2}, \tag{11.16}$$

that yields

$$\alpha(\Lambda^2) = \frac{\alpha(\Lambda_0^2)}{1 - \frac{\alpha(\Lambda_0^2)}{3\pi} \ln(\Lambda^2/\Lambda_0^2)}. \tag{11.17}$$

This has a pole at the momentum scale

$$\Lambda_G = \Lambda_0 e^{\frac{3\pi}{2\alpha}}$$

with a residue that turns out to be negative and is interpreted as implying a negative probability for the occupation of a photon-excited state. Landau considered this an inconsistency of quantum field theory. In practice, $\Lambda_G > \Lambda$, the actual cut-off of QED and at these energies electromagnetism is considered to be already unified with the weak and strong interactions.

Quantum Chromodynamics

The β-function for quantum chromodynamics that deals with quarks and gluons is rather similar to that of QED, but with the opposite sign,

$$\beta_{QCD} = -\frac{21}{12\pi} \alpha^2(\Lambda^2), \tag{11.18}$$

for six quarks flavours (see Huang, 2015). The quantity α here is the analogue of the fine-structure constant for the strong interactions. The negative sign in Eq. 11.18 inverts the flow of the RG equation for the QCD coupling constant α when compared to that of QED. The RG equation 11.14 with $\beta = \beta_{QCD}$ iterates now to weak coupling for large momenta or small distances. This is essentially the phenomenon of asymptotic freedom that we discussed in the context of the non-linear sigma model.

Since the coupling constants of QED and that of the weak interaction grow under renormalisation while that of QCD decreases, this leads to the expectation of a grand unification of these interactions at an energy scale of $\approx 10^{16}$ Gev (Huang, 2015).

In condensed matter there many situations where ghosts can appear and in this case we can fortunately grasp the underlying physics. These situations arise in general as results of a perturbative renormalisation group which has the merit of providing a way to sum a divergent series of powers of logarithms.

Let us consider initially the case of potential scattering by a spinless impurity in an electron gas (Kehrein, 2006). This is described by the Hamiltonian

$$H = \sum_k \epsilon_k c_k^\dagger c_k + g \sum_{k,k'} c_k^\dagger c_{k'}. \tag{11.19}$$

Under a renormalisation group transformation, where electrons are integrated out from the conduction band of width D_0, the renormalisation group equation for the scattering potential, to order g^2 in perturbation theory is given by

$$\frac{dg}{d \ln D} = \rho g^2 \tag{11.20}$$

with a β-function ρg^2. Here, ρ is a constant density of states ($\rho = 1/D_0$) and D is the effective bandwidth at a given step of the renormalisation. This can be integrated to yield

$$g(D) = \frac{g}{1 + \rho g \ln(D_0/D)}. \tag{11.21}$$

Notice that as the effective bandwidth D is reduced under renormalisation, the effective potential becomes weaker for a repulsive potential. However, for an attractive potential it grows and diverges. The divergence in this case is associated with the appearance of a bound state in the system as we know from the exact solution of this problem.

A similar perturbative renormalisation can be carried out for a magnetic impurity in a metal (Kehrein, 2006). This is the Kondo problem, which is described by the Hamiltonian

$$H = \sum_{k\sigma} \epsilon_k c_{k\sigma}^\dagger c_{k\sigma} + J\mathbf{S} \cdot \boldsymbol{\sigma}. \tag{11.22}$$

Eliminating electronic states in the band, one obtains to second order in J the renormalisation group equation

$$\frac{dJ}{d \ln D} = -\rho J^2, \tag{11.23}$$

which integrated leads to

$$J(D) = \frac{J}{1 - \rho J \ln(D_0/D)}. \tag{11.24}$$

It is easy to verify from this solution of the RG equation that for $J < 0$, i.e. when the coupling is ferromagnetic, the system flows to weak coupling as the effective bandwidth is reduced. In this case the localised spin becomes decoupled from the electron gas and the system is asymptotically free. When the coupling is antiferromagnetic ($J > 0$), which is the common situation in metals, the system flows to strong coupling and the coupling even diverges at an energy scale $k_B T_K = D_0 \exp(-1/\rho J)$, with T_K the Kondo temperature. In this case we know the strong coupling solution for this problem, which is the formation of a singlet between the impurity and the conduction electron. The energy scale T_K, at which the effective coupling apparently diverges, marks the crossover from the weak coupling to the strong coupling regimes, physically from a magnetic to a non-magnetic impurity, as we know from the numerical renormalisation group solution (Wilson, 1975).

12

Superconductor Quantum Critical Points

12.1 Introduction

In this chapter we examine a special type of quantum critical point related to superconducting zero-temperature instabilities in many-body systems. We consider the case where the superconductor order parameter is inhomogeneous and characterised by a wave vector q_C. Inhomogeneous ground states also appear in other systems like magnetic materials, in the form of spin-density waves or helicoidal ground states, in charge density wave systems and in excitonic insulators. They share the existence of a characteristic wave vector that determines the spatial modulation of the order parameter. For superconductors, we consider two cases.

First, the problem, treated independently by Fulde and Ferrell (1964) and Larkin and Ovchinnikov (1965), of an s-wave, singlet superconductor in a homogeneous external magnetic field in the absence of vortices or orbital effects. The predicted ground state is generally known as the FFLO superconductor and is described by an order parameter that oscillates in space like $\Delta e^{i q_C \cdot r}$. It arises from the competition between the pairing energy of electrons with opposite spins and the Zeemann energy due to the external magnetic field that forces the alignment of the spins. For sufficiently large magnetic fields the Zeemann energy always wins and the ground state of the system is a spin-polarised, normal fermionic gas. However, as the magnetic field decreases, there is a quantum phase transition to a modulated superconducting state (Samokhin and Marénko, 2006; Caldas and Continentino, 2012) whose nature we identify. We also determine the universality class of this superconducting quantum phase transition. The characteristic wave vector q_C of the resulting modulated superconducting phase is related to the difference between the Fermi wave vectors of the antiparallel spin bands that in turn is proportional to the magnetic field. A related question we also investigate is the fate of the uniform superconductor when the magnetic field is increased in the absence of orbital effects. We show that these two extreme approaches, increasing the field in the homogeneous superconductor and decreasing the field starting from the polarised metal do not merge smoothly into one another.

We also study in this chapter an apparently unrelated problem: that of multi-band superconductivity. This is relevant for many systems such as high-Tc superconductors, heavy fermions, pnictides and other superconducting materials where electrons originating from different atomic orbitals coexist at a common Fermi surface. As before, we consider singlet, s-wave pairing between the electrons. However, in this case, for a two-band system, we have the possibility of pairing electrons in the same band (*intra-band* pairing) or in different bands (*inter-band* pairing).

Suhl, Matthias and Walker (1959) considered an interaction between electrons in different a and b bands (inter-band) of the type

$$\mathcal{H}_{int} = U \sum_{kk'} a_{k\uparrow}^{\dagger} a_{-k\downarrow}^{\dagger} b_{-k'\downarrow} b_{k'\uparrow} + h.c., \tag{12.1}$$

where $a_{k\sigma}^{\dagger}$ $(b_{k\sigma}^{\dagger})$, $a_{k\sigma}$ $(b_{k\sigma})$ are creation and annihilation operators for the electrons in the a (b) band. Decoupling the inter-band pairing above in the usual BCS approximation, the superconducting ground state of the two-band system is characterised by a combination of two intra-band order parameters, $< a_{k\uparrow}^{\dagger} a_{-k\downarrow}^{\dagger} >$ and $< b_{k\uparrow}^{\dagger} b_{-k\downarrow}^{\dagger} >$. This type of inter-band interaction, Eq. 12.1, transfers pairs of electrons, i.e. Cooper pairs, from one band to another. In this sense it represents a *Josephson coupling* between the bands.

Another possibility is to consider an inter-band interaction of the type

$$\mathcal{H}_{int} = U \sum_{kk'} a_{k\uparrow}^{\dagger} b_{-k'\downarrow}^{\dagger} a_{-k\downarrow} b_{k'\uparrow} + h.c. \tag{12.2}$$

In this case a BCS decoupling of the pairing interaction gives rise to inter-band or *hybrid* pairs of the type $< a_{k\uparrow}^{\dagger} b_{-k\downarrow}^{\dagger} >$. This type of pairing should be important in systems with strong inter-band interactions, like heavy fermion systems where the dominant Kondo interaction couples electrons of different bands (the conduction and the narrow f bands). Below, it is shown that this inter-band pairing has a strong tendency to produce modulated superconducting phases with a characteristic wave vector that corresponds to the difference of the Fermi wave vectors of the two interacting bands (a and b). It gives rise to a rich phase diagram and we will concentrate on it.

Let us start with the simplest Hamiltonian describing a two-band system with attractive inter-band interactions

$$\mathcal{H}_0 = \sum_{i,j} t_{ij}^{a} a_i^{\dagger} a_j + \sum_{i,j} t_{ij}^{b} b_i^{\dagger} b_j - U \sum_i n_i^a n_i^b, \tag{12.3}$$

where $n_i^a = a_i^{\dagger} a_i$ and $n_i^b = b_i^{\dagger} b_i$. For simplicity we omitted spin indexes and did not consider intra-band interactions. The bands a and b can be the up and down spin bands of a metal polarised by an external magnetic field, different types of atoms in a cold atom system or the hybridised bands of a multi-band metallic system. The interaction U is a local attractive interaction between fermions in different bands.

In order to study superconductivity in the system described by the Hamiltonian 12.3, we adopt a Green's function approach (see Appendix A.1 on Green's functions). This allows the anomalous Green's function to be obtained, which in turn yields the anomalous correlation function that characterises the superconducting ground state. Also the poles of the Green's functions give the energy of the excitations on this ground state. We use here the well known advanced and retarded Green's functions (Tyablikov, 1967) as defined in Appendix A.1. We start writing the equations of motion for the relevant Green's functions. These are

$$\omega << a_i | a_j^{\dagger} >>_{\omega} = \delta_{ij} + \sum_l t_{il}^a << a_l | a_j^{\dagger} >>_{\omega} - U << a_i b_i^{\dagger} b_i | a_j^{\dagger} >>_{\omega} \quad (12.4)$$

and

$$\omega << b_i^{\dagger} | a_j^{\dagger} >>_{\omega} = - \sum_l t_{il}^b << b_i^{\dagger} | a_j^{\dagger} >>_{\omega} + U << a_i^{\dagger} a_i b_i^{\dagger} | a_j^{\dagger} >>_{\omega} . \quad (12.5)$$

The higher-order Green's functions generated by the dynamics are decoupled as

$$U << a_i^{\dagger} a_i b_i^{\dagger} | a_j^{\dagger} >>_{\omega} \approx -U < a_i^{\dagger} b_i^{\dagger} >< < a_i | a_j^{\dagger} >>_{\omega} = -\Delta << a_i | a_j^{\dagger} >>_{\omega}$$

$$U << a_i b_i^{\dagger} b_i | a_j^{\dagger} >>_{\omega} \approx U < a_i b_i >< < a_i | a_j^{\dagger} >>_{\omega} = -\Delta^* << a_i | a_j^{\dagger} >>_{\omega} .$$

Fourier transforming the decoupled equations, using the translation invariance of the system, which implies that $(U/N) \sum_i < a_i^{\dagger} b_i^{\dagger} > = (U/N) \sum_k < a_k^{\dagger} b_{-k}^{\dagger} > = \Delta$ we obtain a closed system of equations that can be solved to yield

$$<< a_k | a_k^{\dagger} >>_{\omega} = \frac{\omega + \epsilon_k^b}{(\omega - \epsilon_k^a)(\omega + \epsilon_k^b) - |\Delta|^2} \quad (12.6)$$

and, for the *anomalous* Green's function,

$$<< b_{-k}^{\dagger} | a_k^{\dagger} >>_{\omega} = \frac{-\Delta}{(\omega - \epsilon_k^a)(\omega + \epsilon_k^b) - |\Delta|^2}, \quad (12.7)$$

where $\epsilon_k^{a,b} = \sum_l t_{il}^{a,b} \exp[ik \cdot (r_l - r_i)]$. The poles of the propagators in the superconducting state are given by the roots of the equations

$$(\omega - \epsilon_k^a)(\omega + \epsilon_k^b) - |\Delta|^2 = 0.$$

Let us assume that the two bands a and b correspond to the down and up spin bands of a metal in an external magnetic field h, and define

$$\epsilon_k^a = \frac{k^2}{2m} - \mu - h$$

$$\epsilon_k^b = \frac{k^2}{2m} - \mu + h. \tag{12.8}$$

The excitation energy of the quasi-particles, that should be positive for the stability of the superconducting ground state, are given by

$$\omega_\sigma(k) = \sqrt{(\frac{k^2}{2m} - \mu)^2 + |\Delta|^2} - \sigma h. \tag{12.9}$$

The equation for $\sigma = +$ yields a critical magnetic field $h = H_c^M = \Delta$ at which the modes of energy $\omega_+(k)$ become soft for $k = k_F$, i.e. at the Fermi surface of the unpolarised metal. This soft mode is associated with a collapse of the superconducting gap in the whole Fermi surface of the system (Sarma, 1963) and signals an instability of the uniform superconducting ground state. At zero temperature, the gap equation for $h < H_c^M$ can be easily obtained from the anomalous Green's function $<< b_{-k}^\dagger | a_k^\dagger >>_\omega$ using the fluctuation-dissipation theorem (see Appendix). At zero temperature, it is given by

$$\frac{1}{U\rho} = \int_0^\Omega \frac{d\epsilon}{\sqrt{\epsilon^2 + |\Delta|^2}}, \tag{12.10}$$

where ρ is a constant density of states and Ω a high energy cut-off. This equation in the weak coupling regime yields a gap $\Delta = 2\Omega \exp(1/U\rho)$ that is independent of the magnetic field up to a critical field H_c^M where it vanishes abruptly. The discontinuous or first-order character of this zero-temperature transition requires the ground state energies of the different competing phases to be considered and compared. This has been done independently by Chandrasekhar (1962) and Clogston (1962). The competing ground states are the polarised normal metal with energy $F_N = (1/2)\chi_P h^2$ and the superconductor, with a field-independent order parameter Δ that is a solution of Eq. 12.10 at weak coupling, with energy $F_S = (1/2)\rho|\Delta|^2$. Since the Pauli susceptibility of the normal electronic system is $\chi_P = 2\mu_B^2 \rho$, comparing both energies we get a critical field $H_c = (1/\mu_B)|\Delta|/\sqrt{2}$, above which the normal polarised phase has a lower ground state energy than the homogeneous superconductor. This is the actual value of the magnetic field for which the true first-order uniform superconductor-normal quantum phase transition occurs, and is known as the *Chandrasekhar–Clogston* limit. The calculated critical field H_c^M at which soft excitations appear all over the Fermi surface is larger than the true critical field H_c of the first-order transition. This also arises in other discontinuous field induced transitions associated with soft modes (Rezende, 1975, 1977). The critical field H_c^M yields the spinodal that corresponds to the limit of stability of the zero field phase. The true thermodynamic critical field is obtained comparing the energies of the different ground states.

Lines or nodes with gapless excitations are common in superconductors. They can also appear as functions of a control parameter, like magnetic field or pressure due to spin-orbit interactions or k-dependent hybridisations. In this case they are associated with topological transitions which manifest as changes in the order parameter or thermodynamic quantities, like the compressibility, without necessarily destroying the superconducting phase. In the problem considered by Chandrasekhar and Clogston the whole Fermi surface of the system becomes unstable giving rise to the abrupt destruction of the superconducting state.

We started above with a uniform BCS superconductor and applied an external magnetic field to show that as the magnetic field is *increased* to H_C, superconductivity disappears at a first-order quantum phase transition. In the next section we take a different approach. We start with a metal under a sufficiently strong applied magnetic field such that the system is in the normal phase. We then *decrease* the magnetic field and look for superconducting instabilities in the system. At the end of the chapter a consistent view converging from these different approaches is presented.

12.2 Non-Uniform Superconductor

We now calculate the response of the system described by Hamiltonian Eq. 12.3 to a time- and space-dependent *fictitious field* $g(t, q)$ that couples to the superconducting order parameter. The term that describes this coupling and must be added to the Hamiltonian $\mathcal{H}_0$ is given by (Ramires and Continentino, 2010)

$$\mathcal{H}_1 = -g \sum_i e^{iq \cdot r_i} e^{i\omega_0 t} (a_i^\dagger b_i^\dagger + b_i a_i). \tag{12.11}$$

The frequency ω_0 has a small imaginary part to guarantee the adiabatic switching on of the field. We calculate the response of $\mathcal{H}_0$ to first order in perturbation theory in the field g. We assume that the system is in the normal state, such that the superconducting anomalous correlation function $< a_i^\dagger b_i^\dagger > = 0$, in the absence of the field g. We split the Green's functions, normal and anomalous, in two components of zero and first order on the fictitious field g. We write

$$<< a_i^\dagger | b_j^\dagger >> \to << a_i^\dagger | b_j^\dagger >>^{(0)} + << a_i^\dagger | b_j^\dagger >>^{(1)}. \tag{12.12}$$

In the normal phase and in the absence of the fictitious field, the relevant zeroth-order Green's functions can be easily calculated:

$$\ll a_k | a_{k'}^\dagger \gg_\omega^{(0)} = \frac{\delta_{k,k'}}{(\omega - \epsilon_k^a)},$$

$$\ll b_k | b_{k'}^\dagger \gg_\omega^{(0)} = \frac{\delta_{k,k'}}{(\omega - \epsilon_k^b)},$$

$$\ll a_k | b_{k'}^\dagger \gg_\omega^{(0)} = 0,$$

$$\ll a_k^\dagger | a_{k'}^\dagger \gg_\omega^{(0)} = 0,$$

$$\ll b_k^\dagger | b_{k'}^\dagger \gg_\omega^{(0)} = 0,$$

and

$$\ll a_k^\dagger | b_{k'}^\dagger \gg_\omega^{(0)} = \ll b_k^\dagger | a_{k'}^\dagger \gg_\omega^{(0)} = 0.$$

The anomalous first-order Green's functions obey the following equation of motion:

$$i \frac{\partial}{\partial t} << b_i^\dagger(t) | a_j^\dagger(t') >>^{(1)} = - \sum_l t_{il}^b << b_i^\dagger(t) | a_j^\dagger(t') >>^{(1)}$$

$$- U << a_i^\dagger(t) b_i^\dagger(t) a_i(t) | a_j^\dagger(t') >>^{(1)} + g e^{iq \cdot r_i} e^{i\omega_0 t} << a_i(t) | a_j^\dagger(t') >>^{(0)},$$

$$(12.13)$$

where we have neglected terms of second order in the perturbation g and used the previous results for the zeroth-order Green's functions. The decoupling of the higher order Green's function is carried out in the spirit of the random phase approximation (RPA) used before. We have

$$U << a_i^\dagger(t) b_i^\dagger(t) a_i(t) | a_j^\dagger(t') >>^{(1)} \approx \delta \Delta_i(t) << a_i(t) | a_j^\dagger(t') >>^{(0)} . \quad (12.14)$$

In the adiabatic approximation we assume that the time- and space-dependence of the superconducting fluctuations induced by the *fictitious external field* have the same spatial and time dependence of this field, i.e.

$$\delta \Delta_i(t) = < a_i^\dagger(t) b_i^\dagger(t) > = \delta \Delta_q e^{iq \cdot r_i} e^{i\omega_0 t}. \quad (12.15)$$

We proceed Fourier transforming in space the equations of motion. Fourier transforming in time only on the variable t and using the definition

$$a_i^\dagger(t) = (1/2\pi) \int a_i^\dagger(\omega) \exp(-i\omega t) d\omega,$$

we obtain for Eq. 12.13,

$$<< b_k^\dagger | a_{k'}^\dagger(t') >>_\omega^{(1)} = - \frac{(\delta \Delta_q + g)}{\omega + \epsilon_k^b} << a_{k+q} | a_{k'}^\dagger(t') >>_{\omega+\omega_0}^{(0)} . \quad (12.16)$$

Since the first-order Green's function is given in terms of the zeroth-order one, which depends only on the time difference $(t - t')$, the index t' can be made $t' = 0$ or omitted. Using the previous result for $<< a_k | a_{k'}^\dagger >>^{(0)}$, we get

$$<< b_k^\dagger | a_{k'}^\dagger >>_\omega^{(1)} = -\frac{(\delta\Delta_q + g)}{(\omega + \epsilon_k^b)(\omega + \omega_0 - \epsilon_{k+q}^a)}\delta_{k',k+q}. \tag{12.17}$$

Translation invariance has been lost due to the spatial dependence of the fictitious *field*. We can now use the fluctuation-dissipation theorem (see Appendix) to obtain the correlation function $\delta\Delta_q$ from the anomalous Green's functions, we get

$$\delta\Delta_q = \sum_k F_\omega \left\{ \ll b_k^\dagger | a_{k'}^\dagger \gg_\omega^{(1)} \right\},$$

where $F_\omega\{G_k(\omega)\} = -\int d\omega f(\omega)[G_k(\omega + i\epsilon) - G_k(\omega - i\epsilon)]$ is the statistical average of the discontinuity of the Green's functions $G(\omega)$ on the real axis (see Appendix). The function $f(\omega)$ is the Fermi–Dirac distribution.

Introducing a generalised dynamical susceptibility

$$\chi_0(q, \omega_0) = \sum_k F_\omega \left\{ \frac{1}{(\omega + \varepsilon_k^a)} \frac{1}{(\omega + \omega_0 - \varepsilon_{k+q}^b)} \right\}, \tag{12.18}$$

we can write the superconducting response to the *external fictitious field g* in the form

$$\delta\Delta_q(\omega_0) = \frac{\chi_0(q, \omega_0)}{1 - U\chi_0(q, \omega_0)}g. \tag{12.19}$$

The dynamic susceptibility $\chi_0(q, \omega_0)$ can be calculated from Eq. 12.18. It is given by

$$\chi_0(q, \omega_0) = \sum_k \frac{1 - f(\epsilon_{k-q/2}^a) - f(\epsilon_{k+q/2}^b)}{\epsilon_{k-q/2}^a + \epsilon_{k+q/2}^b - \omega_0}. \tag{12.20}$$

12.3 Criterion for Superconductivity

Let us consider Eq. 12.19 in the limit $\omega_0 = 0$. It is clear that when the condition

$$1 - U\chi_0(q, 0) = 0 \tag{12.21}$$

superconductivity can arise in the system even if the fictitious field vanishes (Ramires and Continentino, 2010).

For $q = 0$ and a single-band material with an attractive intra-band interaction U, Eq. 12.21 is the *Thouless criterion* (Thouless, 1960) for the appearance of homogeneous BCS type of superconductivity. For finite temperatures this yields the critical temperature of the BCS superconductor. At zero temperature it gives a critical value for the attractive interaction, $U_c = 0$, such that any finite attractive interaction stabilises the BCS superconductor.

In the case of multi-band systems, or of a single-band material with an applied magnetic field, it may happen that the condition, Eq. 12.21 is first satisfied for a finite $q \neq 0$. In this case, the condition Eq. 12.21 signals the appearance of a q-dependent, modulated superconducting ground state that we call generically an FFLO state (Ramires and Continentino, 2010).

The condition, Eq. 12.21 is formally similar to a generalised Stoner criterion for the appearance of ferromagnetism ($q = 0$) or of a spin-density wave ($q \neq 0$) in metals with repulsive interactions (see Blundell, 2001). In the case of magnetic instabilities we calculate the response of the system to a q-dependent magnetic field. The susceptibility $\chi_0(q, 0)$ is the particle–hole susceptibility instead of the particle–particle one that appears in the superconducting case. Since the former at $T = 0$ reduces essentially to the density of states at the Fermi level $\rho(\epsilon_F)$, the ferromagnetic Stoner criterion is generally expressed as $1 - U\rho(\epsilon_F) = 0$ with U the local Coulomb repulsion.

It is remarkable that in both the attractive and repulsive problems, the interactions between particles of opposite spins are renormalised by fluctuations. In the repulsive magnetic case this is due to particle–particle interactions and has been investigated by Kanamori (1963). It leads to a *saturation* of the repulsive U requiring always a finite density of states for the appearance of magnetism. For attractive interactions, i.e. in the superconducting case, Gorkov and Melik-Barkhudarov (1961) have shown that they get renormalised due to particle–hole corrections which lead, for example, to a reduction of the superconducting critical temperature.

Finally, it calls attention to the fact that, while modulated magnetic ground states are commonly found in nature, modulated superconductors are extremely rare.

12.4 Normal-to-FFLO Quantum Phase Transition in Three Dimensions

Let us consider a one-band metal polarised by an external magnetic field h, such that the a- and b-bands are the up and down spin bands. Their dispersion relations are given by

$$\epsilon_k^{a,b} = \frac{\hbar^2(k^2 - k_F^2)}{2m} \mp h,$$

where k_F is the original Fermi wave vector of the unpolarised band ($h = 0$). The mismatch between the Fermi wave vectors of the up and down spin bands is given by $\delta k_F = k_F^a - k_F^b = h/v_F$, where $v_F = k_F/m$ is the Fermi velocity. At zero temperature, the generalised susceptibility $\chi_0^{ab}(q, \omega_0 = 0)$ can be calculated and we obtain

$$\chi_0^{ab}(q,0) = \rho \left\{ 1 - \ln\left(\frac{h}{v_F k_C}\right) - \frac{1}{2}\left[\frac{1}{\overline{q}}\ln\left(\frac{\overline{q}+1}{\overline{q}-1}\right) + \ln(\overline{q}^2 - 1)\right]\right\}, \quad (12.22)$$

where $\overline{q} = v_F q / 2h = q/2(k_F^a - k_F^b)$, $\rho = 3/(8\pi^3 E_F)$ is the density of states of the unpolarised metal and the Fermi energy $E_F = \hbar^2 k_F^2 / 2m$.

In the normal phase at large magnetic fields, as the field decreases, there is a superconductor quantum critical point (SQCP) at a critical field that separates the normal metal from the non-uniform superconducting phase. This is obtained from the condition, $1 - U\chi_0^{ab}(q,0) = 0$, that can be written as

$$U\rho\left\{1 - \ln\left(\frac{h}{\Delta_0}\right) - \frac{1}{2}\left[\frac{1}{\overline{q}}\ln\left(\frac{\overline{q}+1}{\overline{q}-1}\right) + \ln(\overline{q}^2 - 1)\right]\right\} = 0, \quad (12.23)$$

where $\Delta_0 = v_F k_C \exp(-1/\rho U)$. Notice that a non-uniform superconducting state is only possible for $\overline{q} > 1$. The equation above determines the critical field h_N (or mismatch) for a fixed value of $\overline{q}$. This is given by

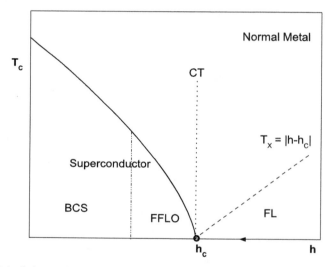

Figure 12.1 Schematic $T_c \times h$ phase diagram for the inter-band superconductor. At zero temperature, as the external magnetic field or Fermi wave vectors mismatch is reduced (indicated by the arrow), the normal metal becomes unstable to an inhomogeneous superconducting phase. This is signalled by the divergence of the q-dependent generalised susceptibility. The FFLO superconductor quantum critical point at h_C has a dynamic exponent $z = 2$ and this allows to determine the behaviour of the thermodynamic quantities along the critical trajectory (CT) as given in the text. As the external field is further reduced there is a first-order transition from the FFLO state to a BCS superconductor (Padilha and Continentino, 2009), which is not studied here. The crossover line $T_\times$ marks the onset of Fermi liquid behaviour with decreasing temperature.

$$\frac{h_N}{\Delta_0} = \frac{e}{(1+\overline{q})} \left(\frac{\overline{q}+1}{\overline{q}-1}\right)^{\frac{\overline{q}-1}{2\overline{q}}}. \tag{12.24}$$

The maximum value of h_N for which the instability occurs is denoted by h_C. It is determined for a wave vector $\overline{q}_C$, which in turn is obtained by differentiating $h_N(\overline{q})$ with respect to $\overline{q}$. From the condition $(\partial \ln h_N / \partial \overline{q})_{\overline{q}=\overline{q}_C} = 0$, we find that

$$\overline{q}_C = \frac{1}{2} \ln \left(\frac{\overline{q}_C + 1}{\overline{q}_C - 1}\right).$$

This gives $\overline{q}_C \cong 1.2$, that substituted in Eq. 12.24 for h_N yields $h_C = h_N(\overline{q}_C) \cong 1.5\Delta_0$. Then, as the field decreases, the normal metal has an instability at the critical field h_C to a modulated superconducting phase characterised by the wave vector $q_C \approx 2.4h/v_F$.

12.5 The Universality Class of the T=0 d=3 FFLO Quantum Phase Transition

Since we are dealing with a quantum phase transition, we know by now that it is essential to consider its dynamics, if we want to characterise its universality class. For this purpose we have to consider the frequency dependence of the particle–particle susceptibility $\chi(q, \omega)$. In the long time scale relevant for critical phenomena the important part of $\chi(q, \omega)$ is that at low frequencies. Expanding Eq. 12.20 for small frequencies ω_0, $h \cong h_C$ and $q \cong q_C$, we get for the denominator of the generalised susceptibility in Eq. 12.19:

$$1 - U\chi_0^{ab}(2q, \omega_0) = U\rho \left[\frac{h - h_C}{h_C} + i\frac{\omega_0}{v_F q_C} + \frac{1}{q_C^2 - 1}(q - q_C)^2\right]. \tag{12.25}$$

The coupling of the fluctuations of the superconductor order parameter to the electronic degrees of freedom gives rise to a characteristic damping of these fluctuations. These modes are purely evanescent with an imaginary dispersion relation.

As in the Hertz (1976) approach to quantum phase transitions, we can construct from the dynamical susceptibility a quantum Gaussian action for this problem. It is written as

$$\mathcal{S} = \int d^d q \int d\omega \left[\delta + \omega + q^2\right] |\psi(q, \omega)|^2, \tag{12.26}$$

where $\delta = h - h_C$ measures the distance to the FFLO SQCP. $\psi(q, \omega)$ is the superconductor order parameter and q above measures the wave vector deviation from q_C. The scaling properties of the free energy associated with this action allows the dynamic exponent of the SQCP at which the FFLO instability occurs to be identified. This is easily found as $z = 2$. Also it immediately identifies the Gaussian correlation length $\nu = 1/2$. The effective dimension of the FFLO quantum phase transition is $d_{eff} = d + z = d + 2$. In three dimensions ($d = 3$), this effective dimension is $d_{eff} = 5$ which is larger then $d_C = 4$, the upper critical dimension for superconducting transitions characterised by a complex order parameter. As a consequence, the Gaussian theory presented above gives the correct description of the FFLO superconducting quantum phase transition in 3d systems. Notice that for $d = 2$ we would also expect that, with additional logarithmic correction (since $d_{eff} = 4$ is the upper critical dimension), the Gaussian approach would also be exact. We show below that this is not the case and that two dimensions is a pathological case. This will be useful as a warning that the simple argument based on the comparison between the effective dimension and the upper critical dimension to determine the universality class of a quantum phase transition may not always work.

What else could go wrong with the Gaussian theory besides the appearance of dangerously irrelevant interactions, as we saw in Chapter 5? Indeed, we did not calculate the coefficient of the quartic term in the effective action. This could change sign from positive to negative, signalling a first-order transition. For $d = 3$, this has been shown not to be the case, but, as we will discuss later, also in this respect, the $d = 2$ may be pathological.

To conclude this section, let us illustrate the power of the scaling approach in making predictions for the $d = 3$ normal–FFLO transition, even for finite temperatures.

Since $d + z > 4$, the quartic interaction in the effective action will turn out to be dangerously irrelevant. As a consequence, it will break down the relation $\psi = \nu z$ between the shift exponent ψ governing the shape of the critical line close to the SQCP. This will be given instead by $\psi = z/(d + z - 2) = 2/3$, as we have shown in previous chapters. The specific heat at the critical field h_c will be dominated by Gaussian fluctuations and given by $C(h = h_C)/T \propto T^{(d-z)/2} = \sqrt{T}$. On the non-critical side of the phase diagram for $h > h_C$, and finite temperatures there will be gapped spin excitations with the gap given by, $\Delta_G \propto (h - h_C)^{\nu z} = (h - h_C)$.

12.6 The Two-Dimensional Problem

Let us now study the above problem in two dimensions (Caldas and Continentino, 2012) where, due to nesting effects, there appear dramatic differences from the the $3d$ case. As an illustration that inhomogeneous superconductivity can arise from other mechanisms besides a magnetic field, we consider the problem of a two-band metal where the mismatch between different Fermi wave vectors is an intrinsic property.

Since we want to avoid problems with the adopted BCS approximation, we consider a two-band, nearly two-dimensional metal with a cylindrical Fermi surface.

We also include a hybridisation between the bands that is very sensitive to the superposition of the different orbitals. Consequently, it can play the role of a control parameter that can be tuned by external pressure.

The Hamiltonian of the two-band system with hybridisation and a purely inter-band interaction is given by

$$\mathcal{H} = H - \sum_{k,\alpha} \mu_\alpha n_\alpha + H_{int} \tag{12.27}$$

$$H = \sum_{k\sigma} \epsilon_k^a a_{k\sigma}^\dagger a_{k\sigma} + \sum_{k\sigma} \epsilon_k^b b_{k\sigma}^\dagger b_{k\sigma} + V \sum_{k\sigma} (a_{k\sigma}^\dagger b_{k\sigma} + b_{k\sigma}^\dagger a_{k\sigma})$$

$$H_{int} = U \sum_{k,k'\sigma} a_{k'\sigma}^\dagger b_{-k'-\sigma}^\dagger b_{-k\sigma} a_{k-\sigma},$$

where ϵ_k^α are the dispersion relations, defined by $\epsilon_k^\alpha = \xi_k^\alpha - \mu_\alpha$, where $\xi_k^\alpha = \frac{\hbar^2 k^2}{2m_\alpha}$ and μ_α is the chemical potential of the (non-interacting) α-particle ($\alpha = a, b$). V is the hybridisation, taken for simplicity as k-independent, which transfers *electrons* between the two bands. Next, as we did before, we calculate the zero temperature response of the two-band system to a wave vector and frequency dependent *fictitious field* that couples to the inter-band superconducting order parameter, $< a_{i\sigma}^\dagger b_{i-\sigma}^\dagger >$. The Hamiltonian associated with this coupling is given by

$$\mathcal{H}_1 = g \sum_i e^{i\mathbf{q} \cdot \mathbf{r}_i} e^{i\omega t} (a_{i\sigma}^\dagger b_{i-\sigma}^\dagger + b_{i\sigma} a_{i-\sigma}), \tag{12.28}$$

where the frequency ω has a small positive imaginary part to guarantee the adiabatic switching on of the fictitious field. As before, we calculate the linear response of the system to the fictitious field g assuming that the hybridisation is too large, such that the system is in the normal phase, i.e. the superconducting order parameter is zero in the absence of the fictitious field.

The superconducting response to the time and q-dependent fictitious field, within the BCS decoupling is obtained as

$$\delta\Delta_q^{ab} = \tilde{\chi}(q, \omega)g = \frac{\chi_V^{12}(q, \omega)}{1 - U\chi_V^{12}(q, \omega)}g, \tag{12.29}$$

where $\delta\Delta_q^{ab} = \sum_k < a_{k+q\sigma}^\dagger b_{-k-\sigma}^\dagger >$ and the non-interacting pair susceptibility $\chi_V^{12}(q, \omega)$ is given by

$$\chi_V^{12}(q, \omega) = \frac{1}{2\pi} \sum_k \frac{1 - f(\epsilon_{k-q}^1) - f(\epsilon_{k+q}^2)}{\epsilon_{k-q}^1 + \epsilon_{k+q}^2 - \omega}. \tag{12.30}$$

The superscripts 1 and 2 refer to the new hybridised bands (see Fig. 12.2)

$$\epsilon_k^{1,2} = \bar{v}_f(k - k_F) \mp V,$$

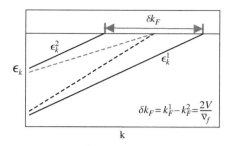

k

Figure 12.2 Simplified band-structure used in the calculation of the dynamic susceptibility. For the original non-hybridised bands (dashed lines) we assume perfect nesting, such that they cross the Fermi surface at the same wave vector. The hybridised bands are shown as full lines and cross the Fermi surface at k_F^1 and k_F^2, with a mismatch $\delta k_F = k_F^1 - k_F^2$. Reprinted figure with permission from Caldas and Continentino (2012), Physical Review B 86, 144503. Copyright 2012 by the American Physical Society.

where $\bar{v}_f = (v_f^a + v_f^b)/2$ is the average Fermi velocity and we took $\mu_\alpha = k_F^2/2m_\alpha$. For simplicity we considered a situation of perfect nesting for the non-hybridised system, such that the two bands cross the Fermi surface at the same k_F for $V = 0$, as shown in Fig. 12.2. The mismatch of the new Fermi wave vectors of the hybridised bands is given by, $\delta k_F = k_F^1 - k_F^2 = 2V/\bar{v}_f$. In the equation for the susceptibility we can neglect spin indexes since the normal system is paramagnetic.

In the normal system, at zero temperature, as the strength of the hybridisation V decreases, the condition $1 - U\chi_V^{12}(q, 0) = 0$ in Eq. 12.29 is eventually satisfied. In this case the interacting pair susceptibility $\tilde{\chi}(q, 0)$ diverges implying that even in the absence of the fictitious field there is the spontaneous appearance of superconductivity in the system.

The nature of the transition

The non-interacting particle–particle, or pair dynamic susceptibility given by Eq. 12.30 can be easily obtained at zero temperature (Caldas and Continentino, 2012). It is given by

$$\chi_V^{12}(q, \omega) = \rho \ln \left[\frac{\frac{2\omega_c}{V}}{1 - \frac{\omega}{2V} + \sqrt{(1 - \frac{\omega}{2V})^2 - \bar{q}^2}} \right] \qquad (12.31)$$

where ρ is the density of states at the Fermi level of the $2d$ system, $\bar{q} = \bar{v}_f q/2V$ and ω_c an energy cut-off. The real part of the static susceptibility is given by

$$\Re e\chi_V^{12}(q,\omega=0) = \rho\left[\ln\frac{2\omega_c}{h} - \ln[1+\sqrt{1-\bar{q}^2}]\right], \bar{q}\leq 1$$

$$= \rho\left[\ln\frac{2\omega_c}{h} - \ln\bar{q}\right], \bar{q}>1. \tag{12.32}$$

The condition for the appearance of superconductivity is given by $1 - U\Re e\chi_{V_c}^{12}(q,0) = 0$ and coincides with that for the divergence of the static interacting pair susceptibility, $\tilde{\chi}(q,\omega=0)$ in Eq. 12.29.

At zero temperature, for a fixed value of the interaction and momentum, as the hybridisation V is reduced, the normal system becomes unstable to a superconducting ground state at a critical value $V = V_c$ determined by the above condition. Using that the superconducting gap for $V = 0$ can be written as $\Delta_0 = 2\omega_c e^{-1/g\rho}$, we obtain for $V_c(\bar{q})$

$$V_c(\bar{q}) = \frac{\Delta_0}{1+\sqrt{1-\bar{q}^2}}, \bar{q}\leq 1$$

$$= \frac{\Delta_0}{\bar{q}}, \bar{q}>1. \tag{12.33}$$

The logarithm of this function is plotted in Fig. 12.3. It is continuous and has a sharp maximum at $\bar{q} = 1$. However, its derivative is discontinuous and different from zero at the maximum.

The value of $\bar{q}$ for which the instability first occurs is that for which $V_c(\bar{q})$ has a maximum, namely, $\bar{q} = \bar{q}_c = 1$. At this value of $\bar{q}$, $V_c = \Delta_0$. Also, using that $k_F^1 - k_F^2 = 2V/\bar{v}_f$, we get $q_c = k_F^1 - k_F^2$, i.e. the wave vector of the instability is exactly that connecting the two Fermi surfaces at k_F^1 and k_F^2, as shown in Fig. 12.4. This turns out to be also true in $1d$, but not in $3d$.

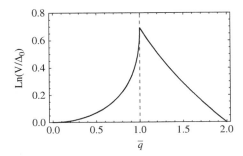

Figure 12.3 The logarithm of the function in Eq. 12.32. This has a maximum for $\bar{q} = 1$ with discontinuous derivatives at this point. Reprinted figure with permission from Caldas and Continentino (2012), Physical Review B 86, 144503. Copyright 2012 by the American Physical Society.

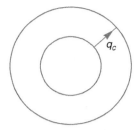

Figure 12.4 The wave vector q_c of the helicoidal superconductor is that joining the two Fermi surfaces at k_F^1 and k_F^2, i.e. $q_c = k_F^1 - k_F^2$ (Caldas and Continentino, 2012).

Next we expand the denominator of the interacting dynamic pair susceptibility, Eq. 12.29, near $V = V_c$, $\bar{q} = \bar{q}_c$, and small frequencies ω. For $q/\bar{q}_c > 1$ and $1 - 2\omega/\bar{v}_f \bar{q}_c < \bar{q}/\bar{q}_c < 1$, we get

$$1 - g\chi_{V_c}^{12}(q, \omega) \approx g\rho \left[\frac{V - V_c}{V_c} + \frac{|\bar{q} - \bar{q}_c|}{\bar{q}_c} + i\sqrt{\frac{|\omega|}{V_c}} \right]. \qquad (12.34)$$

This allows us to write the Gaussian part of the action describing the normal non-uniform superconductor quantum phase transition. It is given by

$$S = \int d^d q \int d\omega \left[\frac{V - V_c}{V_c} + \frac{|q - q_c|}{q_c} + \sqrt{\frac{|\omega|}{V_c}} \right] |\psi_q(q, \omega)|^2. \qquad (12.35)$$

The quantity $(V - V_c)/V_c \sim \delta$ measures the distance to the SQCP associated with the normal-to-FFLO instability at $V_c = \Delta_0$ and with $q_c = 2\Delta_0/\bar{v}_f$. Within this Gaussian approximation, we can define a reduced correlation length $\xi/a = 1/\delta$, where a is the lattice spacing, such that the zero-temperature correlation length exponent takes the value $\nu = 1$ for this model. Also, we can identify that the dynamic critical exponent z that scales the energy, or frequency, at the QCP is given by $z = 2$. Even at the Gaussian level this quantum phase transition is in a new universality class since the critical exponents are non-standard Gaussian exponents. For example, we find $\nu = 1$ instead of the usual Gaussian correlation length exponent $\nu = 1/2$ obtained previously for the $d = 3$ case.

The free energy associated with the Gaussian action in two dimensions, Eq. 12.35, is given by

$$f = -\frac{2}{\pi} \frac{1}{(2\pi)^2} T \int_0^{q_c} dq\, q \int_0^\infty \frac{d\lambda}{e^\lambda - 1} \tan^{-1}\left(\frac{\sqrt{\frac{\lambda T \xi^z}{V_c}}}{1 + q\xi} \right). \qquad (12.36)$$

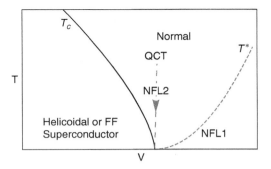

Figure 12.5 Phase diagram of the nearly two-dimensional metal near the normal-to-inhomogeneous superconductor quantum instability at V_c. The critical line $T_c \propto (V - V_c)^{2/3}$. QCT (dashed line) refers to the quantum critical trajectory (see text). The line T^* marks the crossover between two different non-Fermi liquid behaviours, as discussed in the text. The homogeneous superconductor part of the phase diagram for small V and its boundary with the helicoidal or FF phase are not shown in the figure since they are not studied here. Reprinted figure with permission from Caldas and Continentino (2012), Physical Review B 86, 144503. Copyright 2012 by the American Physical Society.

It can be written in the scaling form

$$f \propto |\delta|^{\nu(d+z)} F[\frac{T}{T^*}] \tag{12.37}$$

with $d = 2$, $\nu = 1$ and $z = 2$.

The characteristic temperature $T^* = V_c \xi^{-z} = V_c |\delta|^{\nu z} = V_c |\delta|^2$, in the non-critical side of the phase diagram in Fig. 12.5, corresponds to a crossover temperature between two different regimes. For $T \ll T^*$, the free energy is given by

$$f = \frac{-\zeta(3/2)}{4\pi^{5/2}} \frac{T^{3/2}}{\sqrt{V_c}} [q_c - \xi^{-1} \ln(1 + q_c \xi)]. \tag{12.38}$$

At low temperatures, $T \ll T^*$, the specific heat $C/T = -\partial^2 f/\partial T^2$ has a power law behaviour $C \propto \sqrt{T}$ with a coefficient that increases as the QCP is approached ($\xi^{-1} \propto \delta \rightarrow 0$). So, at low temperatures before the instability, the normal system behaves as a non-Fermi liquid. For $T \gg T^*$, the specific heat besides the $C \propto \sqrt{T}$ term has an additional contribution $C/T \propto \ln T$. The latter is the scaling contribution $C/T \propto T^{(d-z)/z}$ which for $d = z$ gives rise to a logarithmic term.

Now let us assume that the action close to the QCP has an expansion in powers of the order parameter

$$S = \int d\vec{q} \int d\omega \left[\tilde{\chi}^{-1}(q, \omega) \right] |\psi(\vec{q}, \omega)|^2$$

$$+ u \int \prod d\vec{q_i} \int \prod d\omega_i \, \psi(\vec{q_1}, \omega_1)\psi(\vec{q_2}, \omega_2)\psi(\vec{q_3}, \omega_3)$$

$$\times \psi(\vec{q_4}, \omega_4)\delta\left(\sum \vec{q_i}\right)\delta\left(\sum \omega_i\right)$$

with $\tilde{\chi}(q, \omega)$ given in Eq. 12.29. In the second term i varies from 1 to 4 and we assumed a frequency and wave vector independent quartic coefficient u. Simple power counting shows that the correlation function of the order parameter scales as $\psi'^2 = b^{d+z-2+\eta}\psi^2$ and we find the critical exponent $\eta = 1$. Furthermore, the quartic interaction among the fluctuations scales as:

$$u' = b^{2-(d+z)}u, \tag{12.39}$$

such that this interaction is irrelevant in the renormalisation group sense for $d + z \geq 2$. This implies that the Gaussian action gives the correct exponents describing the quantum critical point. However, u is dangerously irrelevant for $d + z > 2$ which leads to departures from naive scaling. For example, the shape of the temperature-dependent critical line shown in Fig. 12.5 is given by $T_c(V) \propto (V_c - V)^\psi$ with the shift exponent $\psi = vz/(1 + v\theta_u)$, where $\theta_u = d + z - 2$ is the scaling exponent of the dangerously irrelevant quartic interaction u_0 (Sachdev, 2011). We obtain $\psi = z/(d + z - 1) = 2/3$, different from the mean field result $\psi = 1/2$ and from the naive scaling prediction $\psi = vz = 2$. The correlation length along the quantum critical trajectory (QCT) in Fig. 12.5 diverges as $T \to 0$ as $\xi \propto (u_0 T^{1/\psi})^{-v} = u_0^{-1}T^{-3/2}$, which makes manifest the dangerously irrelevant character of u. This is different from the naive scaling result $\xi \propto T^{-1/z}$ which does not take into account the dangerously irrelevant nature of u.

We gave above a complete description of the normal-to-inhomogeneous super-conductor quantum phase transition at the Gaussian level. As we have shown, this is the correct theory of this quantum phase transition for a constant u_0 quartic interaction besides corrections due to the dangerously irrelevant nature of this interaction. However, the assumption of taking u constant, i.e. wave vector and frequency independent may not be justified in two dimensions. For this case the frequency-dependence of u is discarded, but its momentum dependence is considered; it has been shown that the quartic interaction obtained from an expansion of the gap equation diverges at the wave-vector $\bar{q}_c = 1$ of the superconducting insta-bility (Shimahara, 1994, 1998; Combescot and Mora, 2005). At finite temperatures this divergence is removed and this also may occur for finite frequencies or moder-ate disorder. A possible consequence of this anomalous behaviour of u is to change the order of the quantum phase transition (Shimahara, 1994, 1998; Combescot and Mora, 2005).

In low-dimensional electronic systems, superconductivity is in general in competition with spin or charge density wave ground states. It is interesting that the correlation length exponent of the incipient $FFLO$ phase, $\nu_{FFLO} = 1$, that we have obtained is larger than that for a competing spin-density wave ($\nu_{SDW} = 1/2$) suggesting that the former superconducting instability should override the latter.

The peculiar quantum critical behaviour of the model above is a consequence that $\bar{q}_c = 1$ is a singular point of the pair susceptibility. In many real systems, we can observe consequences of the structure of Lindhard functions particularly in low-dimensional systems, as the Kohn anomaly and the Peierls instability in one-dimensional conductors.

12.7 Disorder-Induced SQCP

We close this chapter discussing another useful application of the approach presented here for studying SQCP. Let us continue with the $d = 2$ case and consider a nearly two-dimensional, homogeneous superconducting film with disorder. We assume pair breaking disorder caused by magnetic impurities (Ramazashvili and Coleman, 1997) that locally break time reversal invariance, or dissipation due to a coupling of the film to a metallic substrate with electrons tunnelling from one to the other (Mitra, 2008; Deus and Continentino, 2013).

A simple way to treat the effects of disorder or dissipation in the superconductor is to generalise the particle–particle susceptibility to include a finite lifetime of the quasiparticles (Deus and Continentino, 2013). The generalised particle–particle or pair susceptibility $\chi(q, \omega)$ is given by

$$\chi(q, \omega) = \frac{1}{2} \sum_k \left[\frac{\tanh(\beta\epsilon_k/2)}{\epsilon_{k+q} + \epsilon_k - (\omega + i\tau_{SC}^{-1}/2)} + \frac{\tanh(\beta\epsilon_k/2)}{\epsilon_{k-q} + \epsilon_k - (\omega + i\tau_{SC}^{-1}/2)} \right],$$

$$(12.40)$$

where $\beta = (k_B T)^{-1}$. Disorder or dissipation is taken into account by the term $i\tau_{SC}^{-1}$. In the case of impurities, $\tau_{SC} = \Gamma^{-1}$ is the lifetime of the quasi-particles between scattering events.

In the case of a superconducting film deposited on a substrate, their coupling is described by a tunnelling Hamiltonian, $\mathcal{H}_{fs} = t_z \sum_{k_z, k\sigma} a_{k_z, k\sigma}^{\dagger} c_{k\sigma} + h.c.$ We assume the film is in the xy plane, such that there is no momentum conservation in the z direction when an electron in the substrate tunnels to the film and vice versa. The operators a and c refer to electrons in the film and in the bulk metallic substrate, respectively, and k stands for (k_x, k_y). The lifetime of a quasi-particle before tunnelling to the substrate is given by $\tau_{SC} = \Gamma^{-1} \propto t_z^2$.

For $T = 0$ and close to $q = 0$ and $\omega = 0$, we can calculate the pair susceptibility and obtain

$$\chi(q \approx 0, \omega \approx 0) = \rho \left(\frac{1}{4} \ln \left(\frac{\omega_c^4}{\Gamma^4} \right) - \frac{(qv_F)^2}{2\Gamma^2} + i\frac{\omega}{\Gamma} \right), \tag{12.41}$$

where ρ is the constant density of states and ω_c an energy cut-off. Notice that the relevant wave vector to expand is near $q = 0$ since in this case we are interested in the instability of a uniform superconducting state. The Thouless condition $1 - U\Re e \chi(q = 0, \omega = 0) = 0$ yields

$$\frac{1}{U\rho} = \ln \left(\frac{\omega_c}{\Gamma_c} \right), \tag{12.42}$$

or $\Gamma_c = \Omega e^{-\frac{1}{U\rho}} = \Delta_0/2$ for the critical value of disorder below which supercon-ductivity sets in on the film. This disorder induced SQCP can furthermore be fully characterised in the present approach since we know the dynamic and q-dependent pair susceptibility. As before, we can obtain the Gaussian action associated with this problem, which is given by

$$S = \int d^2q \int d\omega \left[\left(\frac{\Gamma - \Gamma_c}{\Gamma_c} \right) + \frac{(qv_F)^2}{2\Gamma^2} + \frac{|\omega|}{\Gamma} \right] |\psi_0(q, \omega)|^2, \tag{12.43}$$

where Γ_c is the critical value of disorder given before. The dynamic exponent turns out to be $z = 2$ and the effective dimension of the QCP, $d_{eff} = d + z = 4$. Then the quantum normal-to-superconductor phase transition in the 2d film is at the upper critical dimension $d_c = 4$ in which case logarithmic corrections to the Gaussian or mean field critical behaviour is expected. It is interesting that near the SQCP, the system is a gapless superconductor (Mitra, 2008), with a density of states, $N(\omega = 0) = \rho/\sqrt{5}$.

As in the previous sections various properties of the thermodynamics and trans-port properties close to the disorder-induced SQCP can be obtained due to the scaling properties of the quantities close to the quantum transition. Also notice that the same action is obtained in three dimensions. Furthermore, if the coefficient of the quartic term due to the interaction between the fluctuations is assumed to be frequency and wave vector independent, it can be shown, as we did before, that it is a dangerously irrelevant perturbation. It will modify the shape of the criti-cal line near the SQCP yielding a shift exponent that depends on dimensionality, $\psi = z/(d + z - 2)$.

13

Topological Quantum Phase Transitions

13.1 The Landau Paradigm

The usual paradigm to describe phase transitions is that of Landau. It is based on an expansion of a free energy functional or an action in terms of an order parameter. Close to the phase transition the order parameter is small, and only terms dictated by symmetry are allowed in this expansion. Associated with this paradigm are many important ideas, such as that of broken symmetry and Goldstone modes. The extension of this paradigm to include quantum phase transitions certainly brought new ideas and progress, as we have seen in this book: the inextricability between the dynamics and static properties, the modified hyperscaling relation with the notion of an effective dimensionality, the role of the uncertainty principle linking energy and temporal fluctuations and so on. The underlying principle here is always that of symmetry. However, as we will see in this chapter, this paradigm is not sufficient to describe certain classes of phenomena that do not involve symmetry changes but are nonetheless still associated with diverging lengths, singularities and critical exponents: features that are clearly related to critical phenomena and phase transitions. We have already met similar behaviour when studying density-driven or Lifshitz transitions. We will see in this chapter that these transitions are much more common and interesting that initially thought. The keyword here is topology, which contains principles and constraints that protect quasi-particle excitations and are still more powerful than those of symmetry.

13.2 Topological Quantum Phase Transitions

We will refer to topological quantum phase transitions (TQPT) as those that separate two phases where at least one of them is topologically non-trivial, as characterised by some topological parameter like a winding number or a Berry phase. In some cases these transitions either give rise to singularities in a thermodynamic quantity or are accompanied by a diverging length. However, differently from Landau-type phase transitions they are not necessarily associated with a symmetry change. Also there is no identifiable order parameter to distinguish between the different phases. Topological quantities are not useful to play the role of an order parameter. They are in general discontinuous at topological transitions and this does not necessarily reflect the nature of the transition. On the other hand, we will show that the renormalisation group is a useful tool to describe TQPT. These transitions can be associated with an unstable fixed point and have critical exponents that obey scaling relations in spite of an unidentified order parameter. They occur at zero temperature and consequently have a quantum character.

13.3 The Kitaev Model

The Kitaev model (Kitaev, 2001, 2003) consists of a linear chain of spinless fermions with an anti-symmetric attractive interaction between fermions in neighbouring sites that gives rise to odd pairing with a p-wave gap. In real space, the Hamiltonian of the model can be written as

$$\mathcal{H} = -\frac{1}{2}\sum_{<ij>} t_{ij} c_i^\dagger c_j - \frac{1}{2}\sum_{<ij>}\left(\Delta_{ij} c_i c_j + \Delta_{ij}^* c_i^\dagger c_j^\dagger\right) - \mu \sum_i c_i^\dagger c_i \qquad (13.1)$$

where $<ij>$ is a nearest neighbour index, t_{ij} is a symmetric nearest neighbour hopping, μ the chemical potential and $\Delta_{ij} = -\Delta_{ji}$ an odd pairing between fermions in neighbouring sites. The operators c_i and $c_i^\dagger$ destroy and create fermions on site i of the chain, respectively. Fourier transforming this Hamiltonian, we obtain

$$\mathcal{H} = \sum_k (\epsilon_k - \mu) c_k^\dagger c_k - \sum_k \left(\Delta_k c_{-k} c_k + \Delta_k^* c_k^\dagger c_{-k}^\dagger\right), \qquad (13.2)$$

where

$$\epsilon_k = -t \cos ka \qquad (13.3)$$

$$\Delta_k = -i \Delta_0 \sin ka, \qquad (13.4)$$

with a the distance between sites on the chain and Δ_0 a constant. This Hamiltonian can also be written in a convenient form to analyse its topological properties:

$$\mathcal{H} = \frac{1}{2}\sum_k \psi_k^\dagger \mathcal{H}_k \psi_k, \qquad (13.5)$$

where $\psi_k^\dagger = (c_k^\dagger, c_{-k})$ and

$$\mathcal{H}_k = \begin{pmatrix} \epsilon_k - \mu & \Delta_k^* \\ \Delta_k & -(\epsilon_k - \mu) \end{pmatrix}. \qquad (13.6)$$

Alternatively, we have

$$\mathcal{H}_k = (\epsilon_k - \mu)\sigma_z + i\Delta_k\sigma_y = -\mathbf{h}(k) \cdot \boldsymbol{\sigma}, \tag{13.7}$$

where we can identify, $h_z(k) = -(\epsilon_k - \mu)$, $h_y(k) = -i\Delta_k$, $h_x(k) = 0$.

The Kitaev Hamiltonian, Eq. 13.1, can easily be diagonalised through a Bogoliubov transformation or in its form in Eq. 13.6, through a direct calculation of the eigenvalues. The energies of the quasi-particles are given by

$$\omega(k) = \pm\sqrt{(-\mu - t\cos ka)^2 + |\Delta_0|^2 \sin^2 ka}. \tag{13.8}$$

Notice these dispersion relations have zero energy modes when $\mu = \pm t$ for $ka = 0$ and $ka = \pi$, i.e. when the chemical potential is at the borders of the conduction band.

The Kitaev model presents two superconducting phases: a weak coupling and a strong coupling superconducting phase. The former is a topologically non-trivial phase and occurs for $|\mu|/t < 1$, i.e. when the chemical potential is inside the non-interacting band of fermions. The strong coupling phase for $|\mu|/t > 1$ is topologically trivial. Both phases are gapped. The gap closes at special k-points for $|\mu|/t = \pm 1$, as given by Eqs. 13.8. At these values of the chemical potential, there are topological quantum phase transitions between the two phases. Since both are superconducting, there is no special symmetry change at the transition. As we pointed out above, Landau approach is of no use here.

The topological nature of the superconducting phases can be characterised by the introduction of a topological invariant, in this case the winding number (Alicea, 2012). Consider the Kitaev model, written in the form $\mathcal{H} = -\mathbf{h}(k) \cdot \boldsymbol{\sigma}$ like in Eq. 13.7. As the wave vector k varies along the one-dimensional Brillouin zone from $k = -\pi/a$ to $k = \pi/a$, the unit vector $\hat{\mathbf{h}}(k) = \mathbf{h}(k)/|\mathbf{h}(k)|$ can describe two types of trajectory depending on the region of the phase diagram, as shown in Fig. 13.1. Since $h_{x,y}(-k) = -h_{x,y}(k)$ and $h_z(-k) = h_z(k)$, we can restrict the values of k to the interval $\pi \geq ka \geq 0$. Defining s_0 and s_π by

$$\hat{\mathbf{h}}(0) = s_0\hat{\mathbf{z}}$$
$$\hat{\mathbf{h}}(\pi/a) = s_\pi\hat{\mathbf{z}}, \tag{13.9}$$

the topological invariant $\mathcal{W} = s_0 s_\pi$ can distinguish the weak coupling and strong coupling phases as it takes the values $\mathcal{W} = -1$ in the former, non-trivial topological phase and $\mathcal{W} = +1$ in the trivial strong coupling one. Notice the geometric meaning of this invariant. The definition $\hat{\mathbf{h}}(k) = \mathbf{h}(k)/|\mathbf{h}(k)|$ maps a trajectory in the Brillouin zone to one in the unit circle. In the non-trivial topological phase as ka varies from 0 to π the unit vector $\hat{\mathbf{h}}$ goes from pointing to one pole of the circle to another. In the trivial phase the unit vector returns to pointing to the same pole

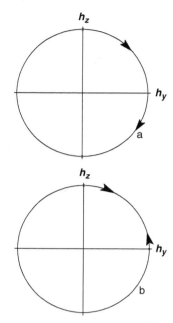

Figure 13.1 The trajectories of the *effective field* $\hat{\mathbf{h}}(k)$ in the unit circle as the wave vector varies from $ka = 0$ to $ka = \pi$, a) in the topological phase; b) in the trivial phase.

as shown in Fig. 13.1 ($s_0 = s_\pi$). In the latter case this is a trivial trajectory that can be collapsed with the initial point, differently from the former. The existence of a topological invariant is a consequence of the symmetries of the Hamiltonian, in this case particle–hole and spinless time reversal symmetry. Notice that the topological invariant can only change sign when the gap in the spectrum of excitations closes, i.e. at the topological quantum phase transition.

The form of the spectrum of excitations near the topological quantum phase transition is given by

$$\omega(k) = \sqrt{(\mu - \mu_c)^2 + |\Delta_0|^2 a^2 k^2}, \qquad (13.10)$$

where $|\mu_c| = t$. At the QCP of the topological transition, $\mu = \mu_c$, the dispersion relation $\omega(k) = |\Delta_0|ak$ is Dirac-like, with a velocity $|\Delta_0|a$. Since the dynamic exponent z is defined by the form of the dispersion relation, $\omega(k) \propto k^z$ at the QCP, the linear spectrum implies that $z = 1$. The critical behaviour of the gap for excitations, $\omega(k = 0) = \Delta_G = (\mu - \mu_c)^{\nu z}$ defines the gap exponent νz. Eq. 13.10 then leads to the value $\nu z = 1$. Since $z = 1$, this in turn allows the correlation length exponent, $\nu = 1$, to be obtained. Further on in the text we will identify the characteristic length associated with the topological quantum phase transition. We will show that this characteristic or correlation length diverges as $\xi \propto (\mu - \mu_c)^{-1}$, yielding an independent confirmation for the result $\nu = 1$.

Properties of the Non-Trivial Topological Phase

The non-trivial topological weak coupling phase of the Kitaev model has properties that make the actual physical implementation of this simple model extremely worthwhile. This phase presents zero energy modes at the extremities of a finite chain with open boundary conditions. These are Majorana fermions which are robust against decoherence due to their physical separation and are associated with the degenerescence of the superconducting ground state in the weak coupling topological phase. Two Majoranas represented by the operators $\alpha_{A,B}$ at site j give rise to a normal fermion such that

$$c_j = \frac{1}{2}(\alpha_{B,j} + i\alpha_{A,j}). \tag{13.11}$$

The anticomutation relations of the fermions c_j require that

$$\{\alpha_{P,j}, \alpha_{P',i}\} = 2\delta_{PP'}\delta_{ij}$$
$$\alpha_{P,j} = \alpha_{P,j}^{\dagger}, \tag{13.12}$$

with P and P' equal A or B. The latter equation, Eq. 13.12, is the defining property of the Majorana fermion, which is its own anti-particle. The Kitaev Hamiltonian written in terms of Majorana operators is

$$\mathcal{H} = -\frac{\mu}{2}\sum_{j=1}^{N}(1 + i\alpha_{B,j}\alpha_{A,j}) - \frac{i}{4}\sum_{j=1}^{N-1}\left[(\Delta_0 + t)\alpha_{B,j}\alpha_{A,j+1} + (\Delta_0 - t)\alpha_{A,j}\alpha_{B,j+1}\right]. \tag{13.13}$$

Now let us consider a point in the phase diagram inside the non-trivial topological phase, say $\mu/t = 0$ and $t = \Delta_0$. In this case the Hamiltonian becomes

$$\mathcal{H} = -i\frac{t}{2}\sum_{j=1}^{N-1}\alpha_{B,j}\alpha_{A,j+1}. \tag{13.14}$$

It can be immediately checked from this equation that the Majoranas at the ends of the chain $\alpha_{A,1}$ and $\alpha_{B,N}$ do not appear in the Hamiltonian and consequently do not contribute to the energy of the system. They give rise to a two-fold degeneracy of the ground state since they can combine to form a normal, although highly non-local, fermion that can be present or not without altering the ground state energy of the system. These zero-energy Majorana modes are a property of the system in its topological phase characterised by the topological invariant $\mathcal{W} = -1$. A nice way to understand these Majorana modes is to think of the vacuum bordering the chain as a trivial superconductor, such that the only way to connect smoothly these phases is by *closing the gap*, which is accomplished by the zero-energy modes at the ends of the chain. As the parameters of the system change inside the topological phase $|\mu/t| < 1$, the Majoranas become less localised but they are always present. They are topologically protected and can disappear only with the closure of the gap at the TQPT at $|\mu/t| = 1$. As we will discuss further on, the penetration length of the Majorana modes inside the chain is the correlation length ξ associated with the TQPT. It diverges at the transition with the critical exponent $\nu = 1$, i.e. $\xi \propto |(\mu/t) - (\mu/t)_c|^{-1}$ where $|(\mu/t)_c| = 1$. This result can be obtained considering the equations of motion for the Majorana operators in the Heisenberg representation, $id\alpha/dt = [\alpha, \mathcal{H}]$. We get

$$\frac{d\alpha_{B,j}}{dt} = (t - \Delta_0)\alpha_{A,j-1} + (t + \Delta_0)\alpha_{A,j+1} - \mu\alpha_{A,j}$$

$$\frac{d\alpha_{A,j}}{dt} = -(t + \Delta_0)\alpha_{B,j-1} - (t - \Delta_0)\alpha_{B,j+1} + \mu\alpha_{A,j}. \tag{13.15}$$

We assume that the time dependence of the Majorana modes is harmonic, i.e. $\alpha_{P,j}(t) = \alpha_{P,j}(t = 0)e^{i\omega t}$ to obtain

$$i\omega\alpha_{B,j} = (t - \Delta_0)\alpha_{A,j-1} + (t + \Delta_0)\alpha_{A,j+1} - \mu\alpha_{A,j}$$

$$i\omega\alpha_{A,j} = -(t + \Delta_0)\alpha_{B,j-1} - (t - \Delta_0)\alpha_{B,j+1} + \mu\alpha_{A,j}. \tag{13.16}$$

For zero-energy modes, the first of Eqs. 13.16 can be written in matrix form as

$$\begin{pmatrix} \alpha_{A,n+1} \\ \alpha_{A,n} \end{pmatrix} = \begin{pmatrix} \frac{\mu}{t+\Delta_0} & \frac{\Delta_0 - t}{\Delta_0 + t} \\ 1 & 0 \end{pmatrix} \begin{pmatrix} \alpha_{A,n} \\ \alpha_{A,n-1} \end{pmatrix} \tag{13.17}$$

and a similar equation for $\alpha_{B,n}$. Obtaining the eigenvalues of the transfer matrix and iterating this equation starting from site 1 to, say, site L, we can obtain the relative amplitude of the Majorana mode at site $L \gg 1$ (DeGottardi, 2011). For $t = \Delta_0$ this can be written as, $\alpha_{A,L} \propto e^{-L/\xi}\alpha_{A,1}$, where $\xi^{-1} = \ln(\mu/t)$ ($\mu/t < 1$). Since $(\mu/t)_c = 1$, we have $\xi = \ln(\mu/t) = \ln[1 + (\mu/t) - (\mu/t)_c]$. Then, sufficiently close to the TQPT, i.e. for $|(\mu/t) - (\mu/t)_c| \ll 1$, we have $\xi^{-1} = |\mu/t - (\mu/t)_c|$. The penetration length of the Majorana mode inside the chain is the characteristic or correlation length of the TQPT and diverges at this transition with the critical exponent $\nu = 1$. Together with the dynamic exponent these yield for the gap exponent, $\nu z = 1$, such that the gap vanishes linearly at the TQPT, i.e. $\Delta_G \propto |\mu/t - (\mu/t)_c|$.

These critical exponents are not independent but are related to at least another exponent through the quantum hyperscaling relation $2 - \alpha = \nu(d + z)$. Here d is the dimensionality of the system, in this case $d = 1$. The exponent α describes the behaviour of the singular part of the free energy density close to the TQPT, i.e.

$f_s \propto |(\mu/t) - (\mu/t)_c|^{2-\alpha}$. Since the compressibility, $\kappa = \partial^2 f_s/\partial\mu^2 \propto |(\mu/t) - (\mu/t)_c|^{-\alpha}$, the exponent α characterises the singularity of the compressibility at the TQPT. Numerical results on the Kitaev chain (Nozadze, 2016) show a logarithmic singularity of the local compressibility at the middle of a long chain at the TQFT at $(\mu/t)_c = 1$. This is consistent with the value $\alpha = 0$ obtained from the quantum hyperscaling relation with $\nu = 1$, $z = 1$ and $d = 1$.

13.4 Renormalisation Group Approach to the Kitaev Model

In the sections above we were able to characterise the critical behaviour of the Kitaev model at its TQPT, without introducing an order parameter as in the Landau approach. Here we show how the renormalisation group without appealing to this concept can also provide a useful approach to TQPT (Continentino *et al.*, 2014).

We consider the Kitaev chain and perform a renormalisation group (RG) transformation that removes every other site in the chain, as we did for the linear chain in Chapter 9. This is a decimation procedure with a length scale factor $b = 2$. It generates a new renormalised chain with a lattice spacing $a' = a/2$. In momentum space this transformation corresponds to double the values of the wave vectors such that $k' = 2k$. Here we apply the renormalisation group transformation in momentum space. In the renormalised lattice k' replaces k in the equations $\epsilon_k = -t\cos ka$ and $\Delta_k = -i\sin ka$, Eqs. 13.3 and 13.4, for the band dispersion relation and the superconducting gap, respectively. Using $k' = 2k$ and the relations,

$$\cos k'a = \cos 2ka = 2\cos^2 ka - 1 \qquad (13.18)$$

$$\sin k'a = \sin 2ka = 2\sin ka\cos ka, \qquad (13.19)$$

we obtain a new Hamiltonian with the same form as the previous one, but with renormalised parameters given by

$$\Omega' = \Omega^2 - 2$$
$$\delta' = 2\delta\sqrt{1 + \delta^2}$$

where $\Omega = -2(\epsilon_k)/t$, $\delta = \Delta_k/\Delta_0$ (Eqs. 13.3 and 13.4). We can write these equations in the form of recursion relations as

$$\Omega_{n+1} = \Omega_n^2 - 2 \qquad (13.20)$$

$$\delta_{n+1} = 2\delta_n\sqrt{1 + \delta_n^2}, \qquad (13.21)$$

where the index n refers to the stage of the renormalisation process. Eq. 13.20 is the logistic map that we have encountered before in Chapter 9. It has two unstable fixed points at $\Omega^* = 2$ and $\Omega^* = -1$. The former divides the Ω axis in two distinct regions: the region $|\Omega| > 2$ where all initial points $|\Omega_0| > 2$ iterate to infinity under successive renormalisation group transformations, and the region $|\Omega| < 2$ where any initial point $|\Omega_0| < 2$ remains always in this interval under iteration. The unstable fixed point at $\Omega^* = 2$ corresponds to the bottom of the band at $ka = 0$. The point $\Omega_0 = -2$ corresponds to the top of the fermion band and iterates to the fixed point $\Omega^* = 2$. The renormalisation group Eq. 13.20 in the region $[-2, 2]$ has periodic orbits but it is *chaotic* for most of the initial points, as any point in this interval is reached arbitrarily close, if the system is iterated a sufficiently large number of times. The fixed point $\Omega^* = -1$ and the point $\Omega_0 = 1$ that maps into $\Omega^* = -1$ correspond to values of k such that $\cos ka = \pm 1/2$, i.e. to $ka = \pm \pi/3$ and $ka = \pm 2\pi/3$.

The recursion relation, Eq. 13.21, has fixed points at $\delta^* = 0$ and $\delta^* = \pm i\sqrt{3/4}$. The former is unstable and corresponds to $ka = 0$ and $ka = \pi$, the bottom and the top of the band of fermions. The latter are also unstable and correspond to $ka = \pm \pi/3$ and $ka = \pm 2\pi/3$. Notice that $\delta_0 = \pm i$ iterate to $\delta = 0$. Initial values of δ between $-i$ and i iterate always in the interval $(-i, i)$ such that the gaps generated by the RG procedure in this region are always smaller than the initial gap Δ_0. For initial points outside this interval δ iterates to $-\infty$.

Summarising, Eqs. 13.20 and 13.21 have an unstable fixed point at the bottom of the conduction band $ka = 0$ corresponding to $(\Omega^*, \delta^*)=(2, 0)$. The states at the top of the band, $ka = \pi$ map to these fixed points. The fixed points $(\Omega^*, \delta^*)=(-1, \pm i\sqrt{3/4})$ correspond to values of $ka = \pm \pi/3$ and $ka = \pm 2\pi/3$ inside the band of conduction states. The iteration of Eqs. 13.20 and 13.21 with initial points in the neighbourhood of these fixed points gives rise to a chaotic sequence where all points in the interval $\Omega \ni [-2, 2]$, $\delta \ni [-i, i]$ are visited arbitrarily close for a sufficient large number of iterations (see Fig. 13.2) for almost all initial values. This rectangle is the attractor of the weak coupling, topologically

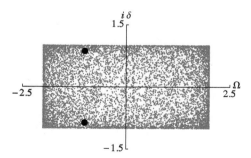

Figure 13.2　The rectangle defined by $\Omega \ni [-2, 2]$, $\delta \ni [-i, i]$ is the attractor of the weak pairing phase. Any initial point in this interval iterates under the RG equations (Eqs. 13.20 and 13.21) to points inside this rectangle. Points outside this interval iterate to the conventional strong coupling attractor at $(\Omega, |\delta|) = (\infty, \infty)$ (Continentino *et al.* (2014), Physics Letters A 378 1561).

non-trivial superconducting phase. Points outside this rectangle iterate to the strong coupling attractor at $(\Omega, |\delta|) = (\infty, \infty)$. Notice that, while the attractor of the weak coupling phase is chaotic, that of the strong coupling phase is a conventional one.

The RG equations as given above consist of two independent equations. Alternatively, we can obtain coupled RG equations with the help of the trigonometric relations. Using Eq. 13.19 we can write the recursion relations as

$$\Omega_{n+1} = \Omega_n^2 - 2 \qquad (13.22)$$

$$\delta_{n+1} = \Omega_n \delta_n. \qquad (13.23)$$

These equations have a structure of fixed points similar to that of Eqs. 13.20 and Eq. 13.21 that we have studied above. There is an unstable fixed point at $\Omega^* = 2, \delta^* = 0$, that correspond to $k = 0$ at the bottom of the fermion band. The fixed points that correspond to $ka = \pi/3$ iterate, such that, if we start with one of them, say $\Omega^* = -1, \delta^* = i\sqrt{3/4}$, we obtain the sequence $(-1, -i\sqrt{3/4})$, $(-1, i\sqrt{3/4})$, $(-1, -i\sqrt{3/4})$, $\cdots$, showing that these fixed points appear now as a periodic orbit with period 2. These states are degenerate as they yield the same values for $|\delta|$. This double degeneracy may be associated with that of the weak pairing phase which in turn is related to the presence of two Majorana fermions at the ends of the chain.

The next periodic orbit is period 4, which iterates to pairs (Ω^*, δ^*) given by $(-1.618, 0.587)$, $(0.618, -0.951)$, $(-1.618, -0.587)$, $(0.618, 0.951)$, $(-1.618, 0.587)$. These states, however, are not degenerate as they give rise to different values for the gap.

From the coupled RG equations, Eqs. 13.22 and 13.23 we can also identify the phases and obtain the critical behaviour of the Kitaev model. When $|\Omega| > 2$, for any $\delta > 0$, the recursion relations, Eqs. 13.22 and 13.23 iterate to the strong coupling fixed point $(\Omega = \infty, |\delta| = \infty)$ which characterises the trivial superconducting strong pairing phase with no special topological properties. Then, the same conventional behaviour appears both in the topological properties and in RG description of this phase.

When Ω lies in the interval $[-2, 2]$ and $|\delta| < 1$, using that at each step of the renormalisation procedure, $\Omega_n/2 = \cos k_n a$ and $(i\Delta_n/\Delta_0) = \sin k_n a$, we can square and add these equations to obtain

$$\Omega_n^2/4 + (i\delta_n)^2 = 1. \qquad (13.24)$$

This is the equation of a unit circle in the plane with axis $\Omega/2$ and $i\delta$ (see Fig. 13.3). Then most of the initial points belonging to this circle (this excludes those points that give rise to periodic orbits or that iterate directly to the fixed points) iterate chaotically but always remain constrained to it. This is shown in Fig. 13.3 and also the period 2 orbit at $\Omega^* = -1, \delta^* = \pm i\sqrt{3/4}$. Also since $|\delta| < 1$, Δ always iterate

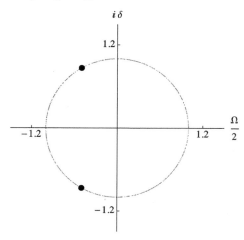

Figure 13.3 Iteration of Eqs. 13.22 and 13.23 starting from $\Omega = \Omega_0$ ($|\Omega_0| < 2$) and $i\delta = \sqrt{1 - \Omega_0^2/4}$. The equation for the attractor where the points iterate is given by $\Omega^2/4 + (i\delta_n)^2 = 1$. The period 2 orbit at $\Omega^* = -1, \delta^* = \pm i\sqrt{3/4}$ is also shown as large dots (Continentino *et al.* (2014), Physics Letters A 378, 1561).

to values smaller than the amplitude Δ_0, whenever Ω lies in the interval $|\Omega| < 2$. Then, the attractor of the weak pairing phase can now be described in terms of a single recursion relation, namely

$$\theta_{n+1} = 2\theta_n, \tag{13.25}$$

modulo 1, with the arguments of the trigonometric function being replace by, $ka \to 2\pi\theta$. Eq. 13.25 is that of the circle map, which has been intensively studied in the theory of chaotic systems (Strogatz, 2014). Notice again that the *weak pairing* phase differently from the usual RG description, as that of the strong pairing phase, is not characterised by an attractive fixed point but by the chaotic attractor shown in Fig. 13.3, and given by Eq. 13.24. Furthermore this curve, Eq. 13.24, contains a periodic orbit of period 2 that we associate with the double degenerescence of this phase. Thus the non-trivial topological properties of the weak pairing phase of the Kitaev model has a counterpart also in its renormalisation group description. Unfortunately, we cannot identify this feature, the chaotic nature of the attractor, unequivocally with a topological non-trivial phase. Indeed, the purely metallic phase of the chain also has a chaotic attractor, that of the logistic map, Eq. 13.22, as we saw in Chapter 9.

The phase diagram of the Kitaev model is shown in Fig. 13.4. As the chemical potential decreases the system goes from the strong pairing to the weak pairing phase and back to strong pairing. The phase transition from the strong to the

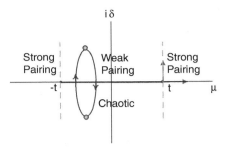

Figure 13.4 The phase diagram of the Kitaev model, showing the weak pairing (chaotic) and strong pairing phases, the unstable fixed points and the flow of the RG equations (arrows) (Continentino *et al.* (2014), Physics Letters A 378, 1561).

weak pairing phase is governed by the unstable fixed point $\Omega^* = 2, \delta^* = 0$. This is the same fixed point that controls the density-driven or Lifshitz transition metal–insulator transition in the non-interacting case. When turning on the pairing interaction, we find it is a relevant perturbation at the fixed point $\Omega^* = 2, \delta^* = 0$. The quantum critical exponents associated with this quantum critical point (QCP) in the presence of interactions can be obtained from the Jacobian of the RG transformations, we get

$$ J = \begin{pmatrix} \frac{\partial \Omega_{n+1}}{\partial \Omega_n} & \frac{\partial \Omega_{n+1}}{\partial \delta_n} \\ \frac{\partial \delta_{n+1}}{\partial \Omega_n} & \frac{\partial \delta_{n+1}}{\partial \delta_n} \end{pmatrix} = \begin{pmatrix} 2\Omega & 0 \\ \delta & \Omega \end{pmatrix}. $$

At the QCP ($\Omega^* = 2, \delta^* = 0$), this yields

$$ J = \begin{pmatrix} 4 & 0 \\ 0 & 2 \end{pmatrix}. $$

Since this is diagonal it implies that the two relevant directions (Ω and δ) are orthogonal. The critical exponents are obtained from the eigenvalues $\lambda_1 = 4$ and $\lambda_2 = 2$. The first, $\lambda_1 = b^z = 4$, yields the dynamic exponent $z = 2$ associated with the quantum critical point of the band-filling or Lifshitz transition for the non-interacting system. The second, $\lambda_2 = 2$, determines the scaling of the interaction $\delta' = 2\delta = b^z \delta$ at the fixed point. This defines the dynamical quantum critical exponent, $z = 1$, for the interacting case. We have made use of the fact above that the scaling factor $b = 2$.

13.5 The Simplest Topological Insulator: the *sp*-Chain

The *sp*-chain has been studied by Schokley (1939) (see also Foo, 1976), who at the time did not dispose of the knowledge of the topological properties of bands in solids. The mystery perceived by the scientists in the 30s now becomes clear in the light of the new concepts that have emerged in recent years (Hasan and Kane, 2010). The sp-chain consists of a one-dimensional atomic system with two orbitals per site of angular momenta $l = 0$ and $l = 1$. Due to their different parities, the hybridisation between these orbitals in neighbouring sites is antisymmetric. This is the crucial ingredient that makes this problem non-trivial. Let us consider an sp-chain described by a two-band Hamiltonian with antisymmetric hybridisation:

$$
\begin{aligned}
\mathcal{H}_{sp} = {} & \epsilon_s^0 \sum_j c_j^\dagger c_j + \epsilon_p^0 \sum_j p_j^\dagger p_j \\
& - \sum_j t_s(c_j^\dagger c_{j+1} + c_{j+1}^\dagger c_j) + \sum_j t_p(p_j^\dagger p_{j+1} + p_{j+1}^\dagger p_j) \\
& + V_{sp} \sum_j (c_j^\dagger p_{j+1} - c_{j+1}^\dagger p_j)) - V_{ps} \sum_j (p_j^\dagger c_{j+1} - p_{j+1}^\dagger c_j), \quad (13.26)
\end{aligned}
$$

where $\epsilon_{s,p}^0$ are the on-site energies of the s and p electrons, $t_{s,p}$ are nearest neighbour hopping matrix elements between electrons in the same orbital, s or p, and the operators $c_j^\dagger$ (c_j), $p_j^\dagger$ (p_j) create (destroy) s and p electrons, respectively, on site j. The hybridisation matrix elements V_{sp} and $V_{ps} = V_{sp}^*$ mix electrons in different orbitals s and p in nearest neighbour sites. Due to the different parities of the orbital states, this hybridisation is odd-parity, such that, $V_{sp}(-x) = -V_{sp}(x)$ or in momentum space $V_{sp}(-k) = -V_{sp}(k)$. Notice that the anti-symmetric nature of this mixing has already been taken into account explicitly in the Hamiltonian. Time reversal symmetry requires that $V_{sp}(k)$ is purely imaginary. However, the mixing term breaks parity symmetry in spite of the fact that the chain is centre-symmetric. We take the chemical potential of the sp-chain, $\mu_{sp} = 0$.

The new bands, or quasi-particles can be obtained introducing, for example, the Greenian operator defined by $\mathbf{G} = (\mathbf{1} - \mathcal{H})^{-1}$.

For the sp model, this is given by

$$
\mathbf{G}_{sp}(k, \omega) = \frac{1}{D_{sp}} \begin{pmatrix} \omega - \epsilon_p^0 - 2t_p \cos ka & 2i\,V_{sp} \sin ka \\ -2i\,V_{sp} \sin ka & \omega - \epsilon_s^0 + 2t_s \cos ka \end{pmatrix}, \quad (13.27)
$$

where

$$
D_{sp}(\omega) = (\omega - \epsilon_s^0 + 2t_s \cos ka)(\omega - \epsilon_p^0 - 2t_p \cos ka) - 4V_{sp}^2 \sin^2 ka. \quad (13.28)
$$

The new hybridised bands are given by the roots of $D_{sp}(\omega) = 0$ and are shown in Fig. 13.5 together with the original non-hybridised bands (Continentino *et al.*, 2014b). We consider the particle–hole symmetric case with $t_s = t_p = t$

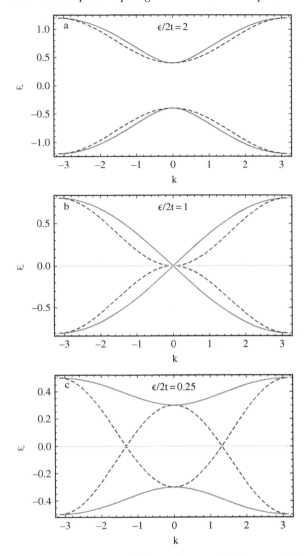

Figure 13.5 Dispersion relations of the 1d sp-model (roots of Eq. 13.28) for $V_{sp} = t$ and different values of the ratio $\epsilon/2t$. a) For $\epsilon/2t > 1$, the system is a standard band insulator with a gap between the valence and conduction bands. In this case the original bands in the absence of hybridisation (dashed lines) do not overlap. b) For $\epsilon/2t = 1$ there is a topological transition between the trivial and non-trivial insulating states. The transition occurs with the closure of the gap between the bands. Near the center of the band, the excitations are Dirac-like fermions with a linear dispersion given by, $\omega = 2V_{sp}k$. c) For $\epsilon/2t < 1$, the system is a topological insulator with zero-energy modes at the ends of a finite chain. Reprinted from Topological states in normal and superconducting p-wave chains, Continentino *et al.*, (2014), Physics Letters A 378, 45-3, 3340, Copyright 2014, with permission from Elsevier.

and $\epsilon_s^0 = -\epsilon_p^0 = \epsilon$. We will show that for $\epsilon/2t = 1$ the sp-chain has a quantum phase transition from a topological insulator to a trivial insulating phase. For $\epsilon/2t > 1$, the system is a standard band insulator with a gap between the valence and conduction bands, as in Fig. 13.5a. For $\epsilon/2t < 1$, the system is a topological insulator with zero energy modes at the ends of a finite chain. The topological quantum phase transition at $\epsilon/2t = 1$ is in the same universality class of that of the Kitaev model with dynamic exponent $z = 1$ and correlation length exponent $\nu = 1$. The former is a direct consequence of the Dirac spectrum at the quantum critical point as shown in Fig. 13.5b. The latter can be obtained analytically, as we did for the Kitaev model, with numerical calculations confirming this result (Continentino *et al.*, 2014b).

The non-trivial topological character of the insulating phase for $\epsilon/2t < 1$ can be shown by writing the Hamiltonian in the Majorana representation (Continentino *et al.* 2014b). This is done writing the fermion operators of the sp-chain Hamiltonian in terms of Majorana fermions. These are given by the following definitions:

$$c_i = \frac{\alpha_{Bi} + i\alpha_{Ai}}{\sqrt{2}}$$

$$p_i = \frac{\beta_{Bi} + i\beta_{Ai}}{\sqrt{2}} \tag{13.29}$$

and similar equations for their complex conjugates, noticing that the Majorana particles are their own antiparticles, i.e. $\alpha_{Ai}^\dagger = \alpha_{Ai}$, $\alpha_{Bi}^\dagger = \alpha_{Bi}$, $\beta_{Ai}^\dagger = \beta_{Ai}$ and $\beta_{Bi}^\dagger = \beta_{Bi}$. In the Majorana basis the Hamiltonian of the sp-chain can be written as (for $t_s = t_p = t$ and $\epsilon_s^0 = -\epsilon_p^0 = \epsilon$)

$$
\begin{aligned}
H_{sp} = {}& i\epsilon \sum_{i=1}^{N} (\alpha_{Bi}\alpha_{Ai} - \beta_{Bi}\beta_{Ai}) \\
& -it \sum_{i=1}^{N-1} (\alpha_{Bi}\alpha_{Ai+i} - \alpha_{Ai}\alpha_{Bi+1}) + it \sum_{i=1}^{N-1} (\beta_{Bi}\beta_{Ai+i} - \beta_{Ai}\beta_{Bi+1}) \\
& +iV \sum_{i=1}^{N-1} (\alpha_{Bi}\beta_{Ai+1} - \beta_{Bi}\alpha_{Ai+1} - \alpha_{Ai}\beta_{Bi+1} + \beta_{Ai}\alpha_{Bi+1}).
\end{aligned} \tag{13.30}
$$

Next we apply a Fourier transformation. For this purpose we introduce

$$\alpha_{Bi} = \sum_k \alpha_k^B e^{ikx_i} \tag{13.31}$$

and similarly for the other Majoranas. Using the properties of the Majorana operators we see that $\alpha_{-k}^B = \alpha_k^{B\dagger}$. We finally get

$$H_{sp} = i \left\{ \sum_k E_k \alpha_k^B \alpha_{-k}^A - \sum_k E_k \beta_k^B \beta_{-k}^A - 2i \sum_k V_{sp}(k) \beta_k^A \alpha_{-k}^B \right.$$

$$\left. +2i \sum_k V_{sp}(k) \beta_k^B \alpha_{-k}^A \right\}, \tag{13.32}$$

where $E_k = \epsilon - 2t \cos ka$ and $V_{sp}(k) = V_{sp} \sin ka$. This Hamiltonian can be written in matrix form in the following way, $H_{sp} = \Psi H \Psi^\dagger$, where

$$H(k) = \begin{pmatrix} 0 & -E_k & 0 & -2i V_{sp}(k) \\ E_k & 0 & 2i V_{sp}(k) & 0 \\ 0 & -2i V_{sp}(k) & 0 & E_k \\ 2i V_{sp}(k) & 0 & -E_k & 0 \end{pmatrix}$$

and $\Psi = (\alpha_k^A, \alpha_k^B, \beta_k^A, \beta_k^B)$. The Majorana number $\mathcal{M}$ is defined as the product of the signs of the Pfaffians of the matrix $H(k)$ at $ka = 0$ and $ka = \pi$. We have $Pf[H(k = 0)] = \epsilon - 2t$ and $Pf[H(ka = \pi)] = \epsilon + 2t$ such that $\mathcal{M} = \text{sgn}(\epsilon - 2t)\text{sgn}(\epsilon + 2t)$. The trivial non-topological phase is characterised by $\mathcal{M} = +1$ while the topological phase is characterised by $\mathcal{M} = -1$. Then, in the case $t_s = t_p = t$ and $\epsilon_s^0 = -\epsilon_p^0 = \epsilon$ and for $\epsilon < 2t$ the sp-chain is a topological insulator and for $\epsilon = 2t$ there is a quantum topological transition from this state to a trivial insulating one as shown in Fig. 13.5b. Notice that, in the derivation above, the criterion for the system to be in the topological phase does not involve directly V_{sp}.

In order to obtain the end modes of the chain, we introduce two new operators (Alexandrov and Coleman, 2014):

$$r_i = c_i + p_i$$
$$l_i = c_i - p_i. \tag{13.33}$$

Using Eqs. 13.29 these can be rewritten in terms of Majorana operators as

$$r_i = \gamma_{Bi}^+ + i\gamma_{Ai}^+$$
$$l_i = \gamma_{Bi}^- + i\gamma_{Ai}^- \tag{13.34}$$

where we introduced new composite or hybrid sp Majorana operators given by

$$\gamma_{A/Bi}^\pm = \alpha_{A/Bi} \pm \beta_{A/Bi}. \tag{13.35}$$

In terms of these hybrid sp Majorana operators, the symmetric ($t_s = t_p = t$, $\epsilon_0^s = -\epsilon_0^p = \epsilon$) sp-chain Hamiltonian can be written as

$$\mathcal{H}_{sp} = i\frac{\epsilon}{2}\sum_{i=1}^{N}(\gamma_{Bi}^{+}\gamma_{Ai}^{-} - \gamma_{Ai}^{+}\gamma_{Bi}^{-}) + 2i(V-t)\sum_{i=1}^{N-1}(\gamma_{Bi}^{-}\gamma_{Ai+1}^{+} - \gamma_{Ai}^{-}\gamma_{Bi+1}^{+})$$

$$-2i(V+t)\sum_{i=1}^{N-1}(\gamma_{Bi}^{+}\gamma_{Ai+1}^{-} - \gamma_{Ai}^{+}\gamma_{Bi+1}^{-}). \tag{13.36}$$

Let us consider the point in the phase diagram $\epsilon = 0$ and $V = t$ inside the topological insulating phase. The Hamiltonian in this case is given by

$$\mathcal{H}_{sp} = -4it\sum_{i=1}^{N-1}(\gamma_{Bi}^{+}\gamma_{Ai+1}^{-} - \gamma_{Ai}^{+}\gamma_{Bi+1}^{-}). \tag{13.37}$$

Notice that the Majorana operators γ_{B1}^{-} and γ_{A1}^{-} do not enter the Hamiltonian so that there are two unpaired hybrid sp Majorana fermions in one end of the chain. The same occurs for γ_{BN}^{+} and γ_{AN}^{+} on the opposite end of the chain. However, two Majoranas in the same extremity can combine to form a single fermionic quasi-particle with hybrid sp character. On the left end we have $l_1 = \gamma_{B1}^{-} + i\gamma_{A1}^{-} = c_1 - p_1$. On the right end we have $r_N = \gamma_{BN}^{+} + i\gamma_{AN}^{+} = c_N + p_N$, a total of two electronic states, one at each end of the chain. These zero energy modes are the protected modes of the topological insulator phase of the particle–hole symmetric sp-chain.

13.6 Weyl Fermions in Superconductors

Up to now we have considered systems where both the topological and trivial phases are gapped and the topological quantum phase transition is necessarily accompanied by the closure of the gap. In this section we consider a p-wave two-band superconductor where the non-trivial topological phase has gapless excitations (Puel *et al.*, 2015). The model describes a spinless two-band system with inter-band superconductivity in 1D, i.e. a chain with two orbitals per site, with angular momenta differing by an odd number, let's say p and s. The pairing between fermions on different bands (inter-band) is always p-wave kind, in the sense that the pairing of spinless fermions is anti-symmetric. The problem can be viewed as a generalisation of Kitaev's model to two orbitals and only inter-band pairing. The Hamiltonian of the system in momentum space is written as

$$\mathcal{H} = \sum_{k}\left\{-\mu\left(c_k^{\dagger}c_k + p_k^{\dagger}p_k\right) + 2t\cos(k)\left(p_k^{\dagger}p_k - c_k^{\dagger}c_k\right)\right.$$

$$\left. -i\Delta\sin(k)\,c_k^{\dagger}p_{-k}^{\dagger}\right\}, \tag{13.38}$$

where μ is the chemical potential and Δ the sp pairing amplitude. Note that the hopping amplitude t has different sign in each band, representing particles for the orbital s and holes for the orbital p. We can write the same Hamiltonian using the Bogoliubov-de Gennes (BdG) representation as

$$\mathcal{H} = \sum_k \mathbf{C}_k^\dagger \mathcal{H}_k \mathbf{C}_k, \tag{13.39}$$

with $\mathbf{C}_k^\dagger = \left(c_k^\dagger\, p_k^\dagger\, c_{-k}\, p_{-k} \right)$ and

$$\mathcal{H}_k = -\mu \Gamma_{z0} - \varepsilon_k \Gamma_{zz} + \Delta_k \Gamma_{yx}, \tag{13.40}$$

where $\Gamma_{ab} = r_a \otimes \tau_b$, $\forall\, a, b = x,\, y,\, z$ and $r_{x,y,z}/\tau_{x,y,z}$ are the Pauli matrices acting on particle-hole/orbitals space, respectively, and $r_0 = \tau_0$ are the 2×2 identity matrix. We have defined $\varepsilon_k = 2t \cos(k)$ and $\Delta_k = \Delta \sin(k)$.

Energy Spectrum

The spectrum of the model Hamiltonian, Eqs. 13.40 and 13.38, can be easily obtained. The dispersion relations of the modes are given by

$$\omega_{1,2}(k) = \varepsilon_k \pm \sqrt{\mu^2 + \Delta_k^2}. \tag{13.41}$$

Looking for gapless points k_0, such that, $\omega_{1,2}(k_0) = 0$ as an indication of topological transitions (Volovik, 2009), we find that the possible solutions satisfy

$$\sin^2 k_0 = \frac{1 - (\mu/2t)^2}{1 + (\Delta/2t)^2}. \tag{13.42}$$

Notice that the system is always gapless whenever $|\mu| \leq 2t$. The existence of these gapless modes or Weyl points represents a substantial difference between this and the Kitaev model; see Fig. 13.6. A solution of the gap equation confirms that both the gapless and the gapped phase for $|\mu| > 2t$ admit stable superconducting solutions. The latter requires a minimum critical value of the pairing interaction to stabilise superconductivity. Deep inside the gapless phase the crossings between bands have a linear dispersion relation (Fig. 13.6a). Furthermore, we note the nodes appear in pairs and disappear only when two nodes are combined, as one can see comparing Figures 13.6a and 13.6b. This is a characteristic of Weyl fermions in 3D or 2D superconductors and in this sense, the model here presented can be called a 1D Weyl superconductor. This Weyl behaviour can be compared to the topological gapless phase that appears between quantum spin Hall and insulating phases in 3D (Murakami, 2007). In this case there are topological monopoles which appear or disappear by creation/annihilation of a pair of a monopole and an anti-monopole.

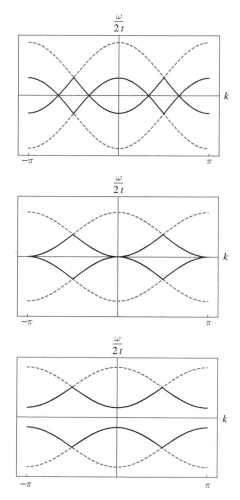

Figure 13.6 Dispersion relations for the sp inter-band superconductor. The system supports gapless excitations when $(|\mu|/2t) \leq 1$ and is gapped otherwise. Specifically, we set a) $(|\mu|/2t) = 0.25$ c) $(|\mu|/2t) = 1.5$. Both regions have self-consistent superconducting solutions (Puel, 2015). The middle panel b) shows the coalescence of the pairs of zero modes at $(|\mu|/2t) = 1$.

Nature of the Transitions

The topological quantum phase transition between the gapless and the gapped superconducting phases by increasing the chemical potential occurs for $\mu/2t = 1$. This transition from the non-trivial topological superconductor, with Weyl fermions, to the topologically trivial one is associated with the collapse of pairs of Weyl points at the centre of the Brillouin zone ($k = 0$) and at its extremities ($k = \pm\pi$), as shown in Fig. 13.6b (Puel, 2015).

Expanding the dispersion relation of the excitations in the gapless phase close to $k = 0$ and $(\mu/2t)_c \approx 1$, we get

$$\omega_2(k) = 2t\sqrt{(1 - \frac{\mu}{2t})^2 + (\frac{\Delta}{2t})^2 k^4}. \tag{13.43}$$

There is also a term proportional to k^2 in this dispersion relation, but we have omitted it since its coefficient is proportional to $(1 - \mu/2t)$ and vanishes at the quantum topological phase transition at $(\mu/2t)_c = 1$. Then, at the quantum critical point, the spectrum of excitations $\omega_2(k) \propto k^2$, which allows the dynamical exponent $z = 2$ to be identified for this transition. Also at $k = 0$, the gap $\omega_2(k = 0) = (\mu/2t)_c - (\mu/2t)$ vanishes linearly at the quantum critical point $(\mu/2t)_c = 1$ with a gap exponent $\nu z = 1$. The critical exponents $\nu = 1/2$ and $z = 2$ show that the quantum phase transition from the topological to the trivial superconducting phase in the inter-band model is in a different universality class from that of the Kitaev model. In the latter at the QCP, $(\mu/2t)_c = 1$, the dispersion is linear, implying a dynamic exponent $z = 1$. These different values of the critical exponents imply distinct behaviour for the compressibility of the two models at the topological quantum phase transition inside the superconducting phase. The compressibility close to this transition is given by $\chi_c = \partial^2 f/\partial\mu^2 \propto |((\mu/2t)_c - (\mu/2t)|^{-\alpha}$ where f is the free energy density. The exponent α is related to the correlation length and dynamical exponents by the quantum hyperscaling relation, $2 - \alpha = \nu(d + z)$. It can be easily verified that while for the intra-band Kitaev model $\alpha = 0$, which is generally associated with a logarithmic singularity, for the inter-band model $\alpha = 1/2$, implying an even stronger singularity for the compressibility at the topological transition. Indeed in the inter-band model the topological transition is in the universality of the Lifshitz transition. Notice that this is a purely topological quantum phase transition, since both phases are characterised by the same order parameter. In spite of this, they have singularities described by critical exponents which obey the quantum hyperscaling relation. The conventional Landau approach of expanding the free energy in terms of an order parameter that becomes small close to a continuous phase transition, and the notion of symmetry breaking, are not useful for this problem. However, the concepts of scaling, universality and critical exponents characterising different singularities still apply, allowing the singularities and temperature dependence of thermodynamic quantities close to the topological phase transition to be predicted.

14

Fluctuation-Induced Quantum Phase Transitions

14.1 Introduction

The quantum phase transitions considered so far in this book are mostly continuous, second order. They are associated with the divergence of a correlation length and of a characteristic time that renders the quantum critical point scale invariant. However, there are many important phase transitions at which the order parameter goes abruptly to zero and, more important, at which the characteristics of length and critical time do not diverge. In general there is also a latent *heat*, or work required, to transform one phase into another. In this and the next chapter we study these discontinuous quantum phase transitions. We proceed by considering initially some important cases of what is known as *fluctuation induced quantum phase transitions* to next treat a more general case of a first-order quantum phase transition. We will show through the study of these problems that the concepts of critical exponents and scaling remain useful, even though these systems do not exhibit scale invariance at their critical point. In Chapter 15 we present the scaling theory of first-order quantum phase transitions and use its results to treat some general problems.

One of the most interesting cases of fluctuation-induced transitions occurs in superconductors due to the coupling of the charged carriers to the electromagnetic field. We consider initially a wider class of systems with macroscopic wave functions, the superfluids, of which superconductors are the charged version. We adopt a macroscopic approach to this problem and describe the superfluid by a complex order parameter φ, such that $|\varphi|^2$ is proportional to the density of condensed particles. The zero temperature phase diagram is expressed in terms of a control parameter, m^2, such that, at $m^2 = 0$, the system presents a quantum phase transition from a superfluid state, for $m^2 < 0$, $\varphi \neq 0$, to a normal state for $m^2 > 0$ where $\varphi = 0$. In the case of the superconductor with particles of charge $q = 2e$, we will include the coupling of these particles to the electromagnetic field and study its effect in the phase diagram of these systems. We show that in the neutral superfluid, the zero temperature transition to the normal state is continuous and occurs at a quantum critical point. However, for the charged superfluid, the superconductor–normal transition becomes discontinuous due to the interaction of the charge carriers with the electromagnetic field. This change in the order of the transition due to the coupling of the order parameter to fluctuations of another field is generally known as *fluctuation-induced first-order transitions*. At finite temperatures the superconducting problem has been investigated by Halperin, Lubensky and Ma (1974). They have shown that the usual, thermal, second-order BCS transition becomes discontinuous due the the coupling of the electrons to the electromagnetic field. These fluctuation-induced phase transitions have also been studied in the context of quantum field theory by Coleman and Weinberg (1973). We will follow the latter approach, which naturally takes into account the quantum effects we are interested in.

The starting point is the Lagrangian density of charged particles minimally coupled to the electromagnetic field. For our purposes the field φ in Eq. 14.1 below is the order parameter of the superfluid (Kleinert, 1987). The Lagrangian density of the superconductor, the electromagnetic field and their coupling is given by

$$L = -\frac{1}{4}(F_{\mu\nu})^2 + (D_\mu\varphi)^\dagger(D_\mu\varphi) - m^2\varphi^\dagger\varphi - \frac{\lambda}{4}(\varphi^\dagger\varphi)^2 \qquad (14.1)$$

where

$$F_{\mu\nu} = \partial_\mu A_\nu - \partial_\nu A_\mu.$$

The first term in the Eq. 14.1 is the Lagrangian density of the electromagnetic field and

$$D_\mu = \partial_\mu - iqA_\mu$$

is the covariant derivative, such that, the two-component scalar φ-field is minimally coupled to the electromagnetic field through the electric charge q. Also, since we work in Euclidean space, $\partial_\mu = (\partial_x, \partial_y, \partial_z, (1/c)\partial_t)$. The superconducting field $\varphi(x, t)$ and the vector potential $A(x, t)$ are space and time dependents. For $m^2 > 0$, the ground state average of the φ-field vanishes, i.e. $\langle\varphi\rangle = 0$. On the other hand for $m^2 < 0$, the ground state average of the field $\langle\varphi\rangle \neq 0$. Then, at $T = 0$, $m = 0$, there is a quantum phase transition from a normal to a superconducting state. When the system is neutral, $q = 0$, the superfluid order parameter decouples from the electromagnetic field and we can think of the φ-dependent part of the Lagrangian as a Landau-type expansion of the energy density of the superfluid.

14.2 Goldstone Modes and Anderson–Higgs Mechanism

Let us neglect for a while the coupling of the φ-field to the electromagnetic field, as for a superfluid of chargeless particles. If we write the two-component scalar field as $\varphi = \rho e^{i\theta}$, the φ part of the Lagrangian density, L_φ, can be written as

$$L_\varphi = \frac{1}{2}(\partial_\mu \rho)^2 + \frac{1}{2}\rho^2(\partial_\mu \theta)^2 - U(\rho)$$

where

$$U(\rho) = m^2 \rho^2 + \frac{\lambda}{4}(\rho^2)^2.$$

The classical solution corresponding to the ground state configuration is obtained by minimising the potential $U(\rho)$; we have:

- $m^2 > 0$, $\rho = 0$
- $m^2 < 0$, $|\varphi|^2 = \rho^2 = -m^2/2\lambda \equiv < \varphi >^2$. We can take $< \theta >= 0$ so that a particular minimum of the potential is chosen. If we write the Lagrangian L_φ in terms of shifted fields, $\rho' = \rho - < \varphi >$ and $\theta' = \theta - < \theta >= \theta$, we get

$$L_\varphi = \frac{1}{2}(\partial_\mu \rho')^2 + \frac{1}{2}(\rho' + < \varphi >)^2(\partial_\mu \theta')^2 - U(\rho' + < \varphi >),$$

which shows that the θ-particle is massless, i.e. $M_\theta = (\partial^2 U/\partial\theta^2)_{\theta=0} = 0$. This zero mass or symmetry restoring mode is the Goldstone mode which restores the original symmetry of the Lagrangian, expressed here by the transformation $\varphi \rightarrow \varphi e^{i\alpha}$, which was broken when we chose the particular ground state with $< \theta >= 0$. Actually, notice that this global gauge symmetry is in fact the symmetry broken by the superfluid state and this plays the role of the rotational symmetry of the Heisenberg model studied in Chapter 2.

Next we consider the charged superfluid or superconductor, $q \neq 0$, with $m^2 < 0$, where symmetry is spontaneously broken. In this case the broken symmetry is a gauge symmetry which takes $\varphi \rightarrow e^{i\alpha(x)}\varphi$ and $A_\mu \rightarrow A_\mu - \partial_\mu \alpha(x)$. As before, the superconducting state is described by the two-component φ-field that we write as, $\varphi = \rho e^{i\theta}$ and $\varphi^\dagger = \rho e^{-i\theta}$. Substituting in the expression for the full Lagrangian density, we get

$$L = -\frac{1}{4}(F_{\mu\nu})^2 + \frac{1}{2}(\partial_\nu \rho)^2 + \frac{1}{2}\rho^2(\partial_\nu \theta + qA_\nu)(\partial_\nu \theta + qA_\nu) - U(\rho^2). \quad (14.2)$$

Defining a new variable $A'_\nu = A_\nu + \frac{1}{q}\partial_\nu \theta$ and expanding the φ field around the minimum, at say, $\rho =< \varphi >$, $< \theta >= 0$, such that, $\rho = \rho' + < \varphi >$, we obtain

$$L = -\frac{1}{4}(\partial_\nu A'_\mu - \partial_\mu A'_\nu)^2 + \frac{1}{2}(\partial_\nu \rho')^2 + \frac{q^2}{2}(\rho' + < \varphi >)^2 A'_\nu A'_\nu - U(\rho' + < \varphi >).$$

From the term $(1/2)q^2 < \varphi >^2 A'_\nu A'_\nu$ of this Lagrangian, we can see that the A'-field has acquired a mass due to the spontaneous symmetry breaking of the φ-field describing the appearance of superconductivity. This is given by

$$m^2_{photon} = \frac{1}{\xi^2_L} = q^2 < \varphi >^2 .$$

This massive photon is essentially at the origin of the Meissner effect in superconductivity and the mass above is just the inverse of the square of the London penetration length ξ_L, $\kappa^2_0 = (1/\xi_L)^2 = q^2 < \varphi >^2$ (Linde, 1979).

This mechanism by which the photon acquires a mass due to its coupling to a field which undergoes a symmetry breaking is known as the *Anderson–Higgs mechanism* (see Negele and Orland, 1988). The existence of a finite plasma frequency in $3d$ is just another manifestation of the same phenomenon (Anderson, 1963).

14.3 The Effective Potential

In this section we use the method of the *effective potential* (Coleman and Weinberg, 1973; Dolan and Jackiw, 1974) to include quantum fluctuations and go beyond the mean field level on our study of the superconducting transition described by the Lagrangian density, Eq. 14.1. This method is described in several textbooks (see for example, Peskin and Schroeder, 1995) and we shall not go into its derivation, but use its results. Essentially, we define a classical field, φ_c, as the vacuum expectation value of the field φ in the presence of a source term coupled to this field. It turns out that the effective action Γ can be expanded in terms of the classical field as

$$\Gamma = \int d^4x [-V_{eff}(\varphi_c) + \frac{1}{2}(\partial_\mu \varphi_c)^2 Z(\varphi_c)]$$

where $Z(\varphi_c)$ is a wave function renormalisation factor, which is unity in the one-loop approximation we shall use, and $V_{eff}(\varphi_c)$ is the *effective potential*. The squared mass of the particles described by the field φ is given in terms of $V_{eff}(\varphi_c)$ as, $m^2 = (d^2 V_{eff}(\varphi_c)/d\varphi_c^2)_{\varphi_c=0}$ and the quartic coupling constant as $\lambda = (d^4 V_{eff}(\varphi_c)/d\varphi_c^4)_{\varphi_c=0}$. Furthermore $Z(0) = 1$.

We call attention that already at this point, an inspection of the time and space derivatives appearing in the Lagrangian density, Eq. 14.1, allows to identify the dynamic critical exponent, $z = 1$, associated with the quantum transition at $g = 0$. We are then dealing again with a Lorentz invariant case in which space and time enter on equal footing in the Lagrangian density. Furthermore, we shall be interested here in the case of spatial dimension $d = 3$, such that, the effective dimension of the problem we are investigating is $d_{eff} = d + z = 4$.

At this point it is useful to rewrite the complex scalar field φ associated with the superconducting state as, $\varphi = \frac{1}{\sqrt{2}}(\varphi_1 + i\varphi_2)$, with φ_1 and φ_2 real. In terms of these fields the Lagrangian density, Eq. 14.1, can be rewritten as

$$L = -\frac{1}{4}(F_{\mu\nu})^2 + \frac{1}{2}(\partial_\mu\varphi_1 + eA_\mu\varphi_2)^2 + \frac{1}{2}(\partial_\mu\varphi_2 - eA_\mu\varphi_1)^2$$
$$-\frac{1}{2}m^2(\varphi_1^2 + \varphi_2^2) - \frac{\lambda}{4!}(\varphi_1^2 + \varphi_2^2)^2. \tag{14.3}$$

The effective potential to be minimised for the problem defined by the Lagrangian density above is given in the one-loop approximation by (Peskin and Schroeder, 1995)

$$V_{eff} = \frac{1}{2}m^2\varphi_c^2 + \frac{\lambda}{4!}\varphi_c^4 + \frac{1}{2}\int \frac{d^4p}{(2\pi)^4} \ln[1 + \frac{\lambda\varphi_c^2/2}{p^2 + m^2}]$$
$$+\frac{1}{2}\int \frac{d^4p}{(2\pi)^4} \ln[1 + \frac{\lambda\varphi_c^2/6}{p^2 + m^2}]$$
$$+\frac{3}{2}\int \frac{d^4p}{(2\pi)^4} \ln[1 + \frac{q^2\varphi_c^2}{p^2 + m^2}] + \frac{1}{2}B\varphi_c^2 + \frac{1}{4!}C\varphi_c^4. \tag{14.4}$$

The first integral in the effective potential comes from the self-coupling term, $(\lambda/4!)\varphi_{c1}^4$. The second comes from the crossed term $2(\lambda/4!)\varphi_{c1}^2\varphi_{c2}^2$. Finally, the third integral comes from the coupling of the charged particles to the electromagnetic field. The factor 3 takes into account the different polarisation of the photons. The last two terms are counter-terms to be fixed in the renormalisation program and which will be used to eliminate the divergencies of the theory.

Let us consider in detail the integrals above. First notice that $d^dp = S_d p^{d-1}dp$ where $S_d = (2\pi)^{d/2}/\Gamma(d/2)$ and $S_4 = 2\pi^2$. Next, we have to deal with the following integral

$$\int_0^\Lambda dx x^3 \ln[1 + \frac{a}{m^2 + x^2}] = \frac{a\Lambda^2}{4} - \frac{m^4 \ln m^2}{4} + \frac{(a + m^2)^2 \ln(a + m^2)}{4}$$
$$+\frac{m^4 \ln(\Lambda^2 + m^2)}{4} - \frac{(a + m^2)^2 \ln(a + m^2 + \Lambda^2)}{4}$$
$$+\frac{\Lambda^4}{4} \ln(1 + \frac{a}{\Lambda^2 + m^2}),$$

which is ultraviolet divergent and requires the introduction of a cut-off Λ. In the limit of very large Λ, we can approximate this integral by

$$\int_0^\Lambda dx x^3 \ln\left[1 + \frac{a}{m^2 + x^2}\right] \approx \frac{a\Lambda^2}{2} + \frac{(a + m^2)^2}{4}\ln(\frac{a + m^2}{\Lambda^2}) - \frac{a^2}{8} - \frac{m^4}{2}\ln\left(\frac{m}{\Lambda}\right)$$

$$\approx \left(\frac{1}{2}a\Lambda^2 + \frac{1}{4}a^2\ln\frac{a}{\Lambda^2} - \frac{1}{8}a^2\right)$$

$$+ \left(\frac{1}{4}a + \frac{1}{2}a\ln\frac{a}{\Lambda^2}\right)m^2 + O(m^4) \tag{14.5}$$

and additional terms that vanish when $\Lambda \to \infty$.

14.4 At the Quantum Critical Point

In the massless case, $m^2 = g = 0$, of the quantum field theory, the system is just at the quantum critical point of the superfluid transition. Substituting the first term in brackets of Eq. 14.5 in the equation for the effective potential we get

$$V_{eff} = \frac{\lambda}{24}\varphi_c^4 + \frac{\lambda\Lambda^2}{48\pi^2}\varphi_c^2 + \frac{3q^2\Lambda^2}{32\pi^2}\varphi_c^2 - \frac{5\lambda^2}{2304\pi^2}\varphi_c^4 - \frac{3q^4}{128\pi^2}\varphi_c^4$$

$$+ \frac{\lambda^2}{256\pi^2}\varphi_c^4\ln(\frac{\lambda\varphi_c^2}{2\Lambda^2}) + \frac{\lambda^2}{2304\pi^2}\varphi_c^4\ln(\frac{\lambda\varphi_c^2}{6\Lambda^2}) + \frac{3q^4}{64\pi^2}\varphi_c^4\ln(\frac{q^2\varphi_c^2}{\Lambda^2})$$

$$+ \frac{1}{2}B\varphi_c^2 + \frac{1}{4!}C\varphi_c^4.$$

The counter-term B is determined, such that, the system remains at the QCP, i.e. that the mass defined as the coefficient of the term $(1/2)\varphi_c^2$ vanishes even when interactions are taken into account. The value of B is then given by

$$B = -\frac{\lambda\Lambda^2}{32\pi^2} - \frac{\lambda\Lambda^2}{96\pi^2} - \frac{3q^2\Lambda^2}{16\pi^2},$$

and the effective potential by

$$V_{eff} = \frac{C}{24}\varphi_c^4 + \frac{\lambda}{24}\varphi_c^4 - \frac{5\lambda^2}{2304\pi^2}\varphi_c^4 - \frac{3q^4}{128\pi^2}\varphi_c^4$$

$$+ \frac{\lambda^2}{256\pi^2}\varphi_c^4\ln\left(\frac{\lambda\varphi_c^2}{2\Lambda^2}\right) + \frac{\lambda^2}{2304\pi^2}\varphi_c^4\ln\left(\frac{\lambda\varphi_c^2}{6\Lambda^2}\right) + \frac{3q^4}{64\pi^2}\varphi_c^4\ln\left(\frac{q^2\varphi_c^2}{\Lambda^2}\right).$$

In order to determine the constant C, we calculate the fourth derivative of V_{eff} with respect to φ_c

$$\frac{d^4V_{eff}}{d\varphi_c^4} = C + \lambda + \frac{55\lambda^2}{144\pi^2} + \frac{33q^4}{8\pi^2} + \frac{11\lambda^2}{32\pi^2}$$

$$+ \frac{9q^4}{8\pi^2}\ln\left(\frac{q^2\varphi_c^2}{\Lambda^2}\right) + \frac{9\lambda^2}{96\pi^2}\ln\left(\frac{\lambda\varphi_c^2}{2\Lambda^2}\right) + \frac{\lambda^2}{96\pi^2}\ln\left(\frac{\lambda\varphi_c^2}{6\Lambda^2}\right).$$

The new coupling constant can not be defined by the value of the 4th derivative at $\varphi_c = 0$ since this is infrared divergent. Consequently, it is defined at an arbitrary value $\varphi_c = M$. From the equation $(d^4 V_{eff}/d\varphi_c^4)_{\varphi_c=M} = \lambda$, we obtain

$$C = -\frac{3\lambda^2}{32\pi^2}\left[\ln\left(\frac{\lambda M^2}{2\Lambda^2}\right) + \frac{11}{3}\right] - \frac{9q^4}{8\pi^2}\left[\ln\left(\frac{q^2 M^2}{\Lambda^2}\right) + \frac{11}{3}\right] - \frac{\lambda^2}{96\pi^2}\left[\ln\left(\frac{\lambda M^2}{6\Lambda^2}\right) + \frac{11}{3}\right].$$

When we substitute this value for C back in the equation for the effective potential, we get

$$V_{eff} = \frac{\lambda}{24}\varphi_c^4 + \left(\frac{5\lambda^2}{1152\pi^2} + \frac{3q^4}{64\pi^2}\right)\varphi_c^4\left[\ln\left(\frac{\varphi_c^2}{M^2}\right) - \frac{25}{6}\right].$$

If we now choose M, which, remember, was arbitrary, to be the minimum of the effective potential above, $M = <\varphi>$, then from the minimum condition $(dV_{eff}/d\varphi_c)_{\varphi_c = <\varphi>} = 0$, we find

$$\lambda = \frac{33}{8\pi^2}q^4, \tag{14.6}$$

where we neglected the term proportional to λ^2 in the effective potential since this is of order q^8. Note that in this equation one of the coupling constants of the problem (λ) has been expressed in terms of the other (q). The final expression for the *Coleman–Weinberg potential* (Coleman and Weinberg, 1973) is

$$V_{eff} = \frac{3q^4}{64\pi^2}\varphi_c^4\left[\ln\left(\frac{\varphi_c^2}{<\varphi>^2}\right) - \frac{1}{2}\right] \tag{14.7}$$

and does not contain λ. This mechanism by which the constant λ has been eliminated is known as *dimensional transmutation*, since we were led to exchange the dimensionless coupling constant λ by the dimensional quantity $<\varphi>$.

In Fig. 14.1 we show the Coleman–Weinberg potential. Notice that this potential has actually a *maximum* at $<\varphi> = 0$. The minima of the potential, Eq. 14.7, occur at *finite values of the classical field*, $\varphi_c = <\varphi> \neq 0$, implying that spontaneous symmetry breaking has taken place. We started at the quantum critical point of the superfluid to discover that symmetry breaking has already taken place. The driving mechanism is the coupling of the electric charge to the electromagnetic potential, the so-called radiative corrections (Coleman and Weinberg, 1973). This phase transition has a completely different nature from those we have considered so far and were associated with the control parameter becoming negative ($m^2 < 0$). In the next section we investigate deeper the nature of this transition.

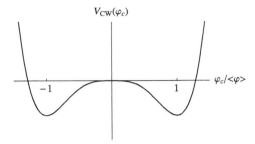

Figure 14.1 The Coleman–Weinberg potential, Eq. 14.7 (Coleman and Weinberg, 1973).

14.5 The Nature of the Transition

We start from the effective potential but now with a finite mass

$$
V_{eff} = \frac{1}{2}m^2\varphi_c^2 + \frac{\lambda}{24}\varphi_c^4 + \frac{\lambda\Lambda^2}{48\pi^2}\varphi_c^2 + \frac{3q^2\Lambda^2}{32\pi^2}\varphi_c^2 - \frac{5\lambda^2}{2304\pi^2}\varphi_c^4 - \frac{3q^4}{128\pi^2}\varphi_c^4
$$
$$
+ \frac{\lambda^2}{256\pi^2}\varphi_c^4 \ln\left(\frac{\lambda\varphi_c^2}{2\Lambda^2}\right) + \frac{\lambda^2}{2304\pi^2}\varphi_c^4 \ln\left(\frac{\lambda\varphi_c^2}{6\Lambda^2}\right) + \frac{3q^4}{64\pi^2}\varphi_c^4 \ln\left(\frac{q^2\varphi_c^2}{\Lambda^2}\right)
$$
$$
+ \frac{m^2}{64\pi^2}\left[\frac{\lambda\varphi_c^2}{2} + \lambda\varphi_c^2 \ln\left(\frac{\lambda\varphi_c^2}{2\Lambda^2}\right)\right] + \frac{3m^2}{32\pi^2}\left[\frac{q^2\varphi_c^2}{2} + q^2\varphi_c^2 \ln\left(\frac{q^2\varphi_c^2}{\Lambda^2}\right)\right]
$$
$$
+ \frac{m^2}{192\pi^2}\left[\frac{\lambda\varphi_c^2}{2} + \lambda\varphi_c^2 \ln\left(\frac{\lambda\varphi_c^2}{6\Lambda^2}\right)\right] + \frac{1}{2}B\varphi_c^2 + \frac{1}{4!}C\varphi_c^4,
$$

where the last three terms proportional to m^2 come from the second term in brackets in Eq. 14.5. We have obtained previously that in the massless case, $m = 0$, the coupling constant $\lambda = O(q^4)$. It will turn out from the lowest-order calculation in λ, i.e. $O(\lambda)$, that $m^2 = O(\lambda)$. Consequently, to first-order in the coupling constant λ, or equivalently to $O(q^4)$ and very near to the QCP, the effective potential is given by

$$
V_{eff} = \frac{1}{2}m^2\varphi_c^2 + \frac{\lambda}{24}\varphi_c^4 + \frac{\lambda\Lambda^2}{48\pi^2}\varphi_c^2 + \frac{3q^2\Lambda^2}{32\pi^2}\varphi_c^2 + \frac{1}{2}B\varphi_c^2
$$
$$
+ \frac{1}{4!}C\varphi_c^4 - \frac{3q^4}{128\pi^2}\varphi_c^4 + \frac{3q^4}{64\pi^2}\varphi_c^4 \ln\left(\frac{q^2\varphi_c^2}{\Lambda^2}\right).
$$

The counter-term B is used to cancel the diverging terms, proportional to the square of the cut-off, by fixing the mass term at the bare mass value m. We get

$$
V_{eff} = \frac{1}{2}m^2\varphi_c^2 + \frac{\lambda}{24}\varphi_c^4 + \frac{1}{4!}C\varphi_c^4 - \frac{3q^4}{128\pi^2}\varphi_c^4 + \frac{3q^4}{64\pi^2}\varphi_c^4 \ln\left(\frac{q^2\varphi_c^2}{\Lambda^2}\right).
$$

From the condition $(d^4 V_{eff}/dx^4)_{x=M} = \lambda$, we obtain

$$C = -\frac{9}{8}\frac{q^4}{\pi^2}\ln q^2 \frac{M^2}{\Lambda^2} - \frac{33}{8}\frac{q^4}{\pi^2}$$

which, when substituted back in the effective potential yields

$$V_{eff} = \frac{1}{2}m^2\varphi_c^2 + \frac{\lambda}{24}\varphi_c^4 + \frac{3q^4}{64\pi^2}\varphi_c^4\left[\ln\left(\frac{\varphi_c^2}{M^2}\right) - \frac{25}{6}\right]. \tag{14.8}$$

Since M is arbitrary, as before, we take it as the minimum of the effective potential $(M =< \varphi >)$, such that $(dV_{eff}/d\varphi_c)_{M=\varphi_c=<\varphi>} = 0$. This condition yields either $< \varphi >= 0$ or $< \varphi > \neq 0$. In the latter case $< \varphi >$ is determined by

$$m^2 + \frac{\lambda}{6} < \varphi >^2 - \frac{11q^4}{16\pi^2} < \varphi >^2 = 0,$$

which yields

$$< \varphi >^2 = \frac{m^2}{(\frac{11q^2}{16\pi^2} - \frac{\lambda}{6})}, \tag{14.9}$$

or alternatively for λ,

$$\lambda = -\frac{6m^2}{< \varphi >^2} + \frac{33q^4}{8\pi^2}. \tag{14.10}$$

This, when substituted back in the effective potential, leads to

$$V_{eff} = \frac{1}{2}m^2\varphi_c^2 - \frac{m^2}{4 < \varphi >^2}\varphi_c^4 + \frac{3q^4}{64\pi^2}\varphi_c^4\left[\ln\left(\frac{\varphi_c^2}{< \varphi >^2}\right) - \frac{1}{2}\right]. \tag{14.11}$$

When the mass vanishes, we get λ as in Eq. 14.6 and the effective potential reduces to the Coleman–Weinberg result. It is now clear that dimensional transmutation can only occur if the effective potential has a non-trivial minimum, i.e. $< \varphi > \neq 0$. The condition for a finite minimum to occur for $m^2 = 0$ is

$$\lambda = \frac{33q^4}{8\pi^2},$$

as can be seen from Eq. 14.9.

An important feature of the problem we are treating is the existence of two characteristic lengths. A coherence or correlation length given by

$$\xi = \frac{1}{\sqrt{2m^2}} \tag{14.12}$$

and another length, similar to the London penetration depth that appears in the Ginzburg–Landau theory of superconductivity. This is defined by

$$\xi_L = \frac{1}{\sqrt{2q^2| <\varphi> |^2}}. \tag{14.13}$$

The ratio of these lengths

$$\frac{\xi_L}{\xi} = \sqrt{\frac{1}{6q^2} \left(\frac{33q^4}{8\pi^2} - \lambda \right)} \tag{14.14}$$

is independent of m^2 and may vanish signalling the divergence of ξ for finite ξ_L. In Ginzburg–Landau theory of superconductivity this ratio determines whether a superconductor is type I or II. Since the results above are obtained for $\lambda \sim q^4$, this would imply $(\xi_L/\xi) \ll 1$ and that they hold only for type I materials, for which $\xi \gg \xi_L$. We will show further on in this chapter, using RG arguments, that this is not necessarily the case.

In Fig. 14.2, we plot the effective potential for different values of the mass m^2. For a *critical value* of the mass, $m^2 = m_c^2$, given by

$$m_c^2 = \frac{3q^4}{32\pi^2} <\varphi>^2 \tag{14.15}$$

there is a first-order phase transition to a new state of broken symmetry with $\varphi_c =< \varphi > \neq 0$. At this value of the mass, the normal and the superconducting phases have the same energy (see Fig. 14.2 for $r = (m^2/m_c^2) = 1$). For $m < m_c$ $(r < 1)$ the true ground state of the system is that with $\varphi_c \neq 0$ and the state with $\varphi_c = 0$ is now metastable, as shown in Fig. 14.2. For $r > 1$ the true ground state is the insulator but the superconductor can exist as a metastable state inside this phase up to $r = 2$, which defines the spinodal or limit of stability of this phase. On the other hand, there is no spinodal in the superconductor where there is always a residue of insulator as a metastable phase.

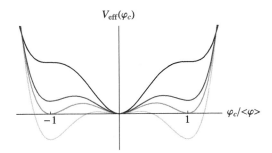

Figure 14.2 The Coleman–Weinberg potential with a finite mass, Eq. 14.11, for different values of $r = (m^2/m_c^2)$. The first order quantum phase transition occurs for $r = 1$ (full black line). Also shown for $r = 0.75$ (dashed line) and $r = 1.25$ (gray line) and for $r = 2$ (dotted line) which is the spinodal and marks the limit of stability of the superconducting phase inside the insulator.

14.6 The Neutral Superfluid

Let us consider the finite temperature phase diagram of the superconductor, as we have been describing here, in terms of a complex scalar field. We start with the neutral superfluid which is decoupled from the electromagnetic field. Using what we learned in previous chapters or carrying out the calculations along the lines described above, we can draw the phase diagram of Fig. 14.3 for $d \geq 3$. The relevant exponents in this case are:

- The shift exponent $\psi = z/(d+z-2) = 1/2$, which describes the critical line of the finite temperature superfluid transitions such that $T_s \propto |g|^{\psi}$. We used $z = 1$, as we have identified previously. The shift exponent remains fixed at the mean field value $\psi = 1/2$ for all $d \geq 3$.
- The crossover exponent $vz = 1/2$ describes the characteristic temperature, $T^* \propto |g|^{vz}$, in the region of the phase diagram for $m > 0$. The nature of the crossover at T^* is discussed below. The dynamic exponent $z = 1$ and for $d => 3$, the correlation length exponent $v = 1/2$ since the effective dimension of the quantum problem is $d + z \geq d_c = 4$, which is the upper critical dimension of the ϕ^4 theory that describes the superfluid. In this case the one loop results describe correctly the quantum phase transition and the critical exponents associated with the quantum critical point at $g = m^2 = 0$ are mean field or Gaussian. Notice that the square of the mass in the quantum field theory is the control parameter g and is related to the correlation length of the statistical problem by

$$\xi = \frac{1}{\sqrt{2m^2}}. \tag{14.16}$$

In the disordered side of the phase diagram, for $m^2 > 0$ and for $T < T^*$, the thermodynamic properties of the system are thermally activated due to the presence of a gap for excitations, $\Delta \propto |g|^{vz} = |g|^{1/2}$. Then in the chargeless case the theory

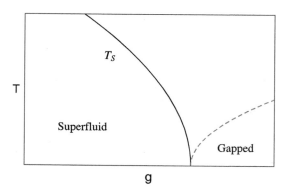

Figure 14.3 The phase diagram of the neutral superfluid. The quantum critical point is at $g = m^2 = 0$.

above describes a quantum superfluid–insulator transition. For $T > T^*$ the free energy is quadratic in temperature as shown below. The $T = 0$ phase transition at $m^2 = 0$ is similar to the interaction driven superfluid–insulator transition studied by Fisher *et al.* (1989), for $d = 3$. These authors started form a microscopic approach to the interacting superfluid to obtain an effective action for their system. The quantum superfluid–insulator transition which they find by varying the interaction among the bosons is in the same universality class of that described by the Lagrangian density above.

14.7 The Charged Superfluid

In order to treat the finite temperature case of the charge superfluid, we note that for quantum theories of Euclidean fields at finite temperatures, the effective potential is equivalent to the thermodynamic free energy (Jackiw, 1973). The generalisation of the effective potential to $T \neq 0$ is done by replacing the frequency integrations by a sum over Matsubara frequencies; in the present case, over bosonic frequencies, i.e.:

$$\int \frac{d^4 p}{(2\pi)^4} \ln[p^2 + m^2 + M^2(\varphi_c)] \rightarrow \frac{1}{\beta} \sum_n \int \frac{d^3 k}{(2\pi)^3} \ln[\vec{k}^2 + \omega_{2n}^2 + M^2(\varphi_c)] \quad (14.17)$$

where $\beta = 1/T$, since we take $k_B = 1$. The effective potential, up to order $O(q^4)$ and finite temperatures, is given by

$$V_{eff}(\varphi_c, T) = \frac{1}{2} m^2 \varphi_c^2 + \frac{\lambda}{24} \varphi_c^4$$
$$+ \frac{3}{2\beta} \sum_n \int \frac{d^3 k}{(2\pi)^3} \ln[\vec{k}^2 + \omega_{2n}^2 + m^2 + q^2 \varphi_c^2].$$

In order to deal with the frequency sums and calculate the temperature dependence of the effective potential, there are a few useful tricks (Belvedere, 1999). We define

$$g(E) = \sum_n \ln[E^2(\vec{k}) + \omega_{2n}^2]$$

but

$$\frac{\partial g(E)}{\partial E} = 2 \sum_n \frac{E}{E^2 + \omega_{2n}^2}.$$

Since, $\omega_{2n}^2 = 2\pi n/\beta$ and

$$\sum_{m=1}^{\infty} \frac{z}{z^2 + n^2} = -\frac{1}{2} z + \coth(\pi z)$$

we get

$$\frac{\partial g(E)}{\partial E} = 2\beta \left(\frac{1}{2} + \frac{1}{e^{\beta E} - 1} \right)$$

and finally

$$g(E) = 2\beta \left(\frac{E}{2} + \frac{1}{\beta} \ln(1 - e^{-\beta E}) \right) + \text{terms independent of } E.$$

Using these results, we obtain

$$\frac{3}{2\beta} \sum_n \int \frac{d^3k}{(2\pi)^3} \ln[\vec{k}^2 + \omega_{2n}^2 + M^2(\varphi_c^2)] = \frac{3}{2} \int \frac{d^3k}{(2\pi)^3} E(\vec{k})$$
$$+ \frac{3}{\beta} \int \frac{d^3k}{(2\pi)^3} \ln(1 - e^{-\beta E(\vec{k})}),$$

$$(14.18)$$

where

$$E^2(\vec{k}) = \vec{k}^2 + M^2(\varphi_c)$$

and,

$$M^2(\varphi_c) = m^2 + q^2 \varphi_c^2.$$

The first term on the right-hand side of Eq. 14.18 is essentially the zero temperature contribution calculated earlier, Eq. 14.8 (Jackiw, 1973), and the second a temperature-dependent term, $\Delta V^1(T, \varphi_c)$, which vanishes at $T = 0$. The integral in the last term can be written as

$$\frac{1}{\beta} \int \frac{d^3k}{(2\pi)^3} \ln(1 - e^{-\beta E(\vec{k})}) = \frac{1}{2\pi^2 \beta^4} I[\beta M(\varphi_c)],$$

where

$$I(y) = \int_0^\infty dx x^2 \ln[1 - e^{-\sqrt{x^2 + y^2}}].$$

Notice that due to the exponential factor, this integral is not ultraviolet divergent. Summing up the results above, we get the temperature-dependent effective potential for the charged superfluid. It is given by

$$V_{eff}(T) = \frac{1}{4} m^2 < \varphi >^2 |g| \left\{ 1 + \frac{2}{\pi^2 m^2 < \varphi >^2 |g|} \frac{T^4}{I} \left(\frac{M(\varphi_c)}{T} \right) \right\}, \quad (14.19)$$

where we defined the distance to the quantum first-order transition, $g = m^2 - m_c^2$, with the critical mass $m_c^2 = (3q^4/32\pi^2) < \varphi >^2$ as given in Eq. 14.15.

We are going to analyse in detail this effective potential in the next chapter after introducing the scaling theory for quantum first-order transitions. Let us anticipate that these equations lead to a line of first-order transitions at finite temperatures given by

$$T_c = \frac{2}{q} \sqrt{m_c^2 - m^2}.$$

For the value of the control parameter corresponding to the quantum critical point of the neutral superfluid, in the case of the superconductor symmetry is now broken at a finite temperature $T_s = (3/\sqrt{8\pi})q < \varphi >$ due to the coupling of the charge carriers to the electromagnetic field. Then, comparing to the neutral superfluid that was massless at its QCP, the charged system has acquired a mass at this point of the phase diagram. This is the Coleman–Weinberg (1973) mechanism of dynamic mass generation.

14.8 Quantum First-Order Transitions in Systems with Competing Order Parameters

Weak first-order transitions and spontaneous symmetry-breaking near quantum phase transitions can also occur due to the coupling of an order parameter to nearly critical fluctuations of another phase. This provides an alternative mechanism for fluctuation-induced quantum first-order transitions besides the coupling to gapless excitations as in the Coleman–Weinberg mechanism or to particle–hole excitations in metals (Belitz *et al.*, 2005a and 2005b). This type of coupling becomes important when two different phases are in competition on the same region of the phase diagram. In this case for some values of the microscopic interactions there are alternative ground states which interfere with one another. We will consider here the case where the competing phases are antiferromagnetism (AF) and superconductivity (SC). We obtain the effects of fluctuations of the order parameter of the competing phase on a given phase in the form of quantum corrections to the effective potential of the latter. We show that these fluctuations can produce spontaneous symmetry breaking and also change the nature of the transition.

In Fig. 14.4 we show the zero temperature phase diagram of a system S close to both an antiferromagnetic and a superconductor quantum critical point. We describe the superconducting phase by a Lorentz invariant-free action, as in the Coleman–Weinberg case, but now coupled to an AF order parameter instead of the electromagnetic field. Later on we consider another free action for the

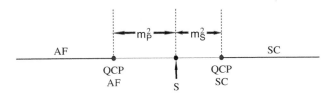

Figure 14.4 A normal, paramagnetic system **S** close to superconducting and antiferromagnetic zero-temperature instabilities.

superconductor which takes into account dissipation and is associated with a
dynamic exponent $z = 2$, instead of $z = 1$ of the Lorentz invariant case
(Ramazashvili and Coleman, 1997; Ramazashvili, 1999).

The classical model is of the Ginzburg–Landau type and contains three real
fields. Two fields, ϕ_1 and ϕ_2, correspond to the two components of the superconductor complex order parameter. The other field, ϕ_3, represents an antiferromagnetic order parameter with a single component, for simplicity. The results can be
generalised to a three-component AF vector field, with the consequence of changing some numerical factors (Ferreira *et al.*, 2004). The free functional associated
with the magnetic part represented by the field ϕ_3, the sub-lattice magnetisation,
takes into account the dissipative nature of the paramagnons near the magnetic
phase transition (Hertz, 1976) in the metal and yields the propagator

$$D_0(\omega, \boldsymbol{q}) = \frac{i}{i|\omega|\tau - q^2 - m_p^2}, \tag{14.20}$$

where τ is a characteristic relaxation time and m_p^2 is the distance to the AF quantum critical point $(AF - QCP)$. The propagator associated with the quadratic,
Lorentz-invariant form of the action of the superconductor is given by

$$G_0(\omega, \mathbf{q}) = \frac{i}{k^2 - m_s^2}, \tag{14.21}$$

where m_s^2 measures the distance to the $SC - QCP$.

The classical potential is given by

$$\begin{aligned}
V_{cl}(\phi_1, \phi_2, \phi_3) = {} & \frac{1}{2}m^2(\phi_1^2 + \phi_2^2) + \frac{1}{2}m_p^2\phi_3^2 \\
& + V_s(\phi_1, \phi_2) + V_p(\phi_3) + V_i(\phi_1, \phi_2, \phi_3),
\end{aligned} \tag{14.22}$$

where the self-interaction of the superconductor field is

$$V_s(\phi_1, \phi_2) = \frac{\lambda}{4!}(\phi_1^2 + \phi_2^2)^2, \tag{14.23}$$

and that of the antiferromagnet,

$$V_p(\phi_3) = \frac{g}{4!}\phi_3^4. \tag{14.24}$$

Finally, the last term is the (minimum) interaction between the relevant fields:

$$V_i(\phi_1, \phi_2, \phi_3) = u(\phi_1^2 + \phi_2^2)\phi_3^2. \tag{14.25}$$

This term is the first allowed by symmetry on a series expansion of the interaction. Notice that for $u > 0$, which is the case here, superconductivity and
antiferromagnetism are in competition and this term does not break any symmetry of the original model. However, including quantum fluctuations, we show that

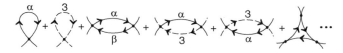

Figure 14.5 One-loop diagrams contributing to the effective potential.

spontaneous symmetry breaking can occur in the normal phase separating the SC and AF phases. This case is represented schematically in Fig. 14.4 and is discussed next.

The first quantum correction to the potential can be obtained by the summation of all one-loop diagrams (Fig. 14.5). It is possible to generalise the method proposed by Coleman (Coleman, 1988) with minimum modifications to account for the different nature of the propagators in this problem (Ferreira *et al.*, 2004). The sum over the field indices can be easily done, if we define a vertex matrix $\mathbf{M}$ given by

$$[M]_{lm} = -i K_0^l \frac{\partial^2 V_{cl}}{\partial \phi_l \partial \phi_m}\bigg|_{\{\phi\}=\{\phi_c\}} \qquad (14.26)$$

and then take the trace. In Eq. (14.26) the propagators ($K_0^l = G_0$ or D_0) are incorporated in the definition of the matrix. We draw the loops with arrows and choose the outgoing propagator of each vertex to be included in the associated element. The matrix $\mathbf{M}$ is then obtained deriving the classical potential with respect to the fields $\{\phi\}$ and taking the values of these derivatives at the classical values of the fields, $\{\phi_{ic}\}$. The sum of diagrams with the correct Wick factors is formally done in momentum space. Using the property of the trace

$$\mathrm{Tr}[\ln(1 - M)] = \ln \det[(1 - M)], \qquad (14.27)$$

we get

$$V^{(1)}[\phi_c] = \frac{i}{2} \hbar \int d^4 k \ln \det [1 - M(k)] . \qquad (14.28)$$

The 3×3 matrix $\mathbf{M}$ can be simplified if we choose the classical minimum of the superconductor fields imposing $\phi_{2c} = 0$ (this can be done because the minimum depends only on the modulus $\phi_{1c}^2 + \phi_{2c}^2$). Hence, rotating to Euclidean space, so that, $k^2 = \omega^2 + q^2$ and using $\hbar = 1$ units the first quantum correction can be written as

$$V^{(1)}(\phi_{1c}, \phi_{3c}) = \frac{1}{2} \int \frac{d^4 k}{(2\pi)^4} \left\{ \ln \left(1 + \frac{A(\phi_{1c}, \phi_{3c})}{k^2 + m^2} \right) \right.$$
$$+ \ln \left[\left(1 + \frac{B(\phi_{1c}, \phi_{3c})}{k^2 + m^2} \right) \left(1 + \frac{C(\phi_{1c}, \phi_{3c})}{|\omega|\tau + q^2 + m_p^2} \right) \right.$$
$$\left. \left. - \left(\frac{D^2(\phi_{1c}, \phi_{3c})}{(k^2 + m^2)(|\omega|\tau + q^2 + m_p^2)} \right) \right] \right\}, \qquad (14.29)$$

where

$$A(\phi_{1c}, \phi_{3c}) = (\lambda/6)\phi_{1c}^2 + 2u\phi_{3c}^2 \tag{14.30}$$

$$B(\phi_{1c}, \phi_{3c}) = (\lambda/2)\phi_{1c}^2 + 2u\phi_{3c}^2 \tag{14.31}$$

$$C(\phi_{1c}, \phi_{3c}) = 2u\phi_{1c}^2 + (g/2)\phi_{3c}^2 \tag{14.32}$$

$$D(\phi_{1c}, \phi_{3c}) = 4u\phi_{1c}\phi_{3c}. \tag{14.33}$$

The total effective potential with first-order quantum corrections is then given by

$$V_{eff}(\phi_{1c}, \phi_{3c}) = V_{cl}(\phi_{1c}, \phi_{3c}) + V^{(1)}(\phi_{1c}, \phi_{3c}) \tag{14.34}$$

where V_{cl} is the classical potential of Eq. (14.22) and $V^{(1)}$ is the first quantum correction of order $\hbar$ of Eq. (14.29).

14.9 Superconducting Transition

We first consider the effects of the antiferromagnetic fluctuations on the supercon-ductor transition. Detailed calculations of the effective potential using $\phi_{3c} = 0$ in Eq. 14.29 yield (Ferreira *et al.*, 2004),

$$V_{eff}(\phi_c) \approx \frac{1}{2}m_s^2\phi_{1c}^2 + \frac{\lambda}{4!}\phi_{1c}^4 + am_p^2\phi_{1c}^2|\phi_{1c}| + \mathcal{O}(\phi^5), \tag{14.35}$$

where m_s^2 and λ are new renormalised quantities, the latter of the same order of the bare original coupling λ. Terms proportional to λ^2 and $m_s^2\lambda$ have been neglected and terms of higher order than m_p^2 also. These are all supposed to be small. The new coupling $a \sim u^{3/2}$ due to the interaction with the AF fluctuations can produce a symmetry breaking in the normal state, extending the SC region in the phase diagram at $T = 0$ towards the $AF - QCP$, but in the paramagnetic phase. It also can turn the transition to weak first order with a small latent heat (Continentino and Ferreira, 2007).

As in the case of the superconductor coupled to the electromagnetic field, we introduce two characteristic lengths, ξ the correlation length and ξ_L, which are given by

$$\xi = \sqrt{1/2|m_s^2|} \tag{14.36}$$

$$\xi_L = \sqrt{1/u|\langle\phi_1\rangle^2|}. \tag{14.37}$$

The condition for the existence of minima $< \phi_1 >$, away from the origin, of the potential in Eq. 14.35 is $\xi_L \ll \xi$, as in the previous case.

Since the first-order transition is produced by the cubic term in Eq. 14.35 and this term is proportional to the *magnetic mass* m_p^2, then at the $AF - QCP$ ($m_p^2 = 0$)

the only effect of this coupling would appear as a term proportional to ϕ^5. This term is usually neglected since its power is higher than those initially considered in the classical potential and usually insufficient to create new minima. Then, at the $AF - QCP$ the effects of the quantum corrections due to the superconducting transition can be neglected.

14.10 Antiferromagnetic Transition

We are interested now in the effect of the superconducting fluctuations in the antiferromagnetic transition. Therefore we look for a partial symmetry broken phase with $\phi_{1c} = \phi_{2c} = 0$ and $\phi_{3c} \neq 0$. Taking this in Eq. 14.29 and considering the Lorentz invariant propagator of the superconducting fluctuations, the effective potential is obtained as (Ferreira *et al.*, 2004)

$$V_{eff}(\phi_{3c}) \approx \frac{1}{2}m_p^2\phi_{3c}^2 + \frac{g}{4!}\phi_{3c}^4 + u^2\phi_{3c}^4 \ln\left(\frac{\phi_{3c}^2}{\langle\phi_3\rangle^2}\right) + um_s^2\phi_{3c}^2 \ln\left(\frac{\phi_{3c}^2}{\langle\phi_3\rangle^2}\right),$$

(14.38)

where m_p^2 and the coupling constants g and u have been renormalised. Notice from Eq. (14.38) that if the superconductor order parameter fluctuations were critical, i.e. $m_s^2 = 0$, we would obtain an effective potential analogous to that of the Coleman–Weinberg case. Away, but close to the $SC - QCP$, the last term of Eq. 14.38 linear in u leads to new and interesting changes in the ground state, as shown in Fig. 14.6. We obtain, besides the two finite minima of the Coleman–Weinberg type potential, two extra minima very close to the origin (Continentino and Ferreira, 2007), i.e., associated with small values of the antiferromagnetic order parameter. This *small moment antiferromagnetic phase (SMAF)* exchanges stability with a *large moment antiferromagnetic phase (LMAF)* giving rise to an additional first order transition when compared to the case with $m_s^2 = 0$. This *SMAF–LMAF* transition takes place even before the continuous antiferromagnetic mean field transition. When the system moves away from the $SC - QCP$ or approaches the $AF - QCP$, i.e. m_p^2 is reduced, the system goes from a paramagnetic phase to a *SMAF* as shown in Fig. 14.6 and then to a *LMAF* phase. The *SMAF* phase is always the closest to the superconductor.

Now, for many cases of interest, SC fluctuations are better described by a dissipative propagator associated with a $z = 2$ dynamics (Ramazashvili and Coleman, 1997). This is useful to account for pair breaking interactions as magnetic impurities that can destroy superconductivity (Mineev and Sigrist, 2001). It is given by

$$G_0(\omega, \boldsymbol{q}) = \frac{i}{i|\omega|\tau' - q^2 - m_s^2}.$$

(14.39)

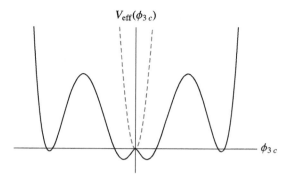

Figure 14.6 New minima appear in the potential for $um_s^2 \neq 0$. Dashed line $m_s^2 = 0$. Full line $um_s^2 \neq 0$.

The parameter m_s^2 is still related with the distance to the $SC - QCP$ and we have introduced a relaxation time τ'. In general, we also have a non-dissipative term (as in the case studied above) but this term is neglected since the linear term is the most important in the low frequency region (Mineev and Sigrist, 2001). Calculation of the effective potential is very similar to the previous cases and the result has the form

$$V_{eff} = \frac{1}{2}m_p^2\phi_3^2 + \frac{1}{4!}g\phi_3^4 + \frac{1}{15\pi^2}(2u\phi_3^2 + m^2)^{5/2} \tag{14.40}$$

with a renormalised mass m_p^2 and coupling constant g. Quantum corrections can again produce a weak first-order transition. The transformation $(\phi_3')^2 = 2u\phi_3^2 + m^2$ makes the potential of Eq. (14.40) simpler and the analysis of its extrema shows that the transition can be first order depending on the coupling values. The appearance of a $SMAF$ phase is not possible in this case.

14.11 One-Loop Effective Potentials and Renormalisation Group

In the previous sections we have obtained weak quantum first-order transitions from minima-generated balancing terms in the effective potential. This is possible when we compare terms of the same order and therefore the minima depend on the values of the couplings. Considering, for example, the superconductor coupled to the electromagnetic field, the terms we have to balance are proportional to λ and q^4. New minima away from the origin are produced if

$$\lambda \sim q^4 \tag{14.41}$$

and this condition also leads to a relation between the London penetration depth and the coherence length, namely $\lambda_L \ll \xi$. In this section we show how we can use renormalisation group arguments to generalise the results for any small λ and q and get rid of the condition (14.41).

When deriving Eq. 14.8 we introduced a new parameter M associated with a renormalisation mass, which was completely arbitrary. The results we obtain must be the same for any chosen value of M. If we prove that a variation in the value of M produces a correspondent variation in the coupling constant λ, it is always possible to satisfy the condition 14.41 choosing a suitable value of M. This can be proven considering Eq. 14.8. We want to rewrite this equation with a value M', such that

$$V_{eff} = \frac{1}{2}m^2\varphi^2 + \frac{\lambda}{4!}\varphi^4 + \frac{3q^4}{64\pi^2}\varphi^4\left[\ln\left(\frac{\varphi^2}{M'^2}\frac{M'^2}{M^2}\right) - \frac{25}{6}\right]$$

$$V_{eff} = \frac{1}{2}m^2\varphi^2 + \frac{\lambda}{4!}\varphi^4 + \frac{3q^4}{64\pi^2}\varphi^4\ln\frac{M'^2}{M^2}$$
$$+ \frac{3q^4}{64\pi^2}\varphi^4\left[\ln\left(\frac{\varphi^2}{M'^2}\right) - \frac{25}{6}\right]$$

which is equivalent to Eq. 14.8 with the reparametrisation

$$\lambda' = \lambda + \frac{9q^4}{8\pi^2}\ln\frac{M'^2}{M^2}. \qquad (14.42)$$

In this sense the effective potential is always given by the same equation with a suitable parametrisation of the coupling.

This argument can be formally stated in a renormalisation group approach and we refer to the correspondent section of Coleman (Coleman, 1988). As a result, a weak first-order transition occurs for any small values of λ and q, since we can appropriately choose the renormalisation mass. The only restriction comes from perturbation theory which requires small couplings.

For the case of coupling between AF and SC order parameters we can also construct a renormalisation group. Within this approach we expect to find all the conditions for the occurrence of weak first-order transitions.

14.12 Conclusions

In this chapter we have studied the effects on a quantum phase transition due to the coupling the order parameter of this phase to soft modes or to fluctuations of a competing phase. These effects are more important at low temperatures and for this reason we have mostly considered the zero-temperature case. The effects of quantum corrections due to this coupling were calculated using the effective potential method. These corrections can produce radical modifications on the nature of the original phase transition, changing it from continuous to discontinuous. They can also affect the ordered phase itself, giving rise to an inhomogeneous phase with two values of the order parameter.

An important and common feature of the problems studied in this chapter is the existence of two length scales. In the next chapter we will present a scaling approach to first-order quantum phase transitions. This will be specially interesting in the case of the weak first-order transitions studied here where the interplay of the two length scales gives rise to interesting effects.

15

Scaling Theory of First-Order Quantum Phase Transitions

15.1 Scaling Theory of First-Order Quantum Phase Transitions

As we have seen in several chapters of this book, scaling theories are invaluable tools in the theory of quantum critical phenomena. They yield relations among critical exponents and the scaling forms in which thermodynamic variables appear in different scaling functions. In the study of strongly correlated metals close to a quantum instability they led to the discovery of a new characteristic temperature T_{coh} that marks the onset of Fermi liquid behaviour. At zero temperature, the scaling theory relies on scale invariance close to a quantum critical point where the characteristic length and time diverge. Here we study the extension of scaling ideas to quantum first-order phase transitions (Continentino and Ferreira, 2004). Although there is no diverging length or time in these transitions, scaling ideas have proved to be very useful for discontinuous, temperature-driven transitions (Fisher and Berker, 1982; Nienhuis and Naunberg, 1975) and will turn out to be also for the case of first-order quantum phase transitions.

Let us consider the scaling form of the singular part of the ground state energy density close to a quantum phase transition

$$f \propto |g|^{2-\alpha}, \tag{15.1}$$

where g measures the distance to the transition that occurs at $g = 0$. The exponent α is related to the correlation exponent ν through the quantum hyperscaling relation $2 - \alpha = \nu(d + z)$, where d is the dimension of the system and z the dynamic critical exponent. The total internal energy density close to the transition can be written as

$$U(g = 0^{\pm}) = U(g = 0) \pm A_{\pm}|g|^{2-\alpha} \tag{15.2}$$

for $g \to 0^{\pm}$. Then, the existence of a first-order phase transition at $T = 0$ with a discontinuity in dU/dg and a *latent heat* implies for the critical exponent α, the value $\alpha = 1$. If quantum hyperscaling applies, this leads to a correlation length exponent $\nu = 1/(d + z)$. This is the quantum equivalent of the classical result $\nu = 1/d$ for temperature-driven first-order transitions (Fisher and Berker, 1982). Associated with this value of the correlation length there is on the disordered side of the phase diagram a new energy scale, $T^* \propto |g|^{z/(d+z)}$.

The presence of a discontinuity in the order parameter and the assumption of no-decay of its correlation function imply $\beta = 0$, as in the classical case (Fisher and Berker, 1982) and $d + z - 2 + \eta = 0$, respectively. As for classical first-order transitions $\delta = \infty$, and for consistency with the scaling relations the order parameter susceptibility seems to diverge with an exponent $\gamma = 1$ (Fisher and Berker, 1982).

If the quantum transition is driven by pressure, $g \propto (P - P_c)/P_c$ where P_c is the critical pressure and a finite *latent heat* means in this case a finite amount of work, $W = A_+ + A_- = P_c \Delta V$, required to bring one phase into another. Such finite *latent work* is associated with a change in volume since the intensive variable, pressure in this case, remains constant at the transition. In the case of a density-driven first-order transition the chemical potential remains fixed while the number of particles changes. Another important constraint on pressure-induced first-order quantum phase transitions is imposed by thermodynamics. The Clausius–Clapeyron equation, $dT_c/dP = \Delta V/\Delta S$ implies $dT_c/dP \to \infty$ as $T \to 0$ since, according to the third law of thermodynamics, the difference in entropy between the two phases must vanish as $T \to 0$.

In the next sections we study two problems that present first-order quantum transitions and confirm the results obtained above on the basis of scaling arguments. The study of specific problems is important to clarify the meaning and range of application of a scaling analysis in situations where criticality is in fact avoided.

15.2 The Charged Superfluid and the Coleman–Weinberg Potential: Scaling Approach

In the previous chapter we obtained the temperature-dependent Coleman–Weinberg effective potential for a charged superfluid. For the case of an arbitrary dimension d, this is given by

$$V_{eff}(T) = \frac{1}{4}m^2\langle\varphi\rangle^2|g|\left[1 + \frac{2}{\pi^2 m^2\langle\varphi\rangle^2}\frac{T^{d+1}}{|g|}I_d\left(\frac{M(\varphi)}{T}\right)\right], \tag{15.3}$$

where the integral I_d is given by

$$I_d(y) = \int_0^\infty dx\, x^{d-1}\ln\left[1 - e^{-\sqrt{x^2+y^2}}\right] \tag{15.4}$$

and $M^2(\varphi) = m^2 + q^2 <\varphi>^2$.

The finite temperature phase diagram associated with this potential is shown in Fig. 15.1. For completeness, we show also the critical line of the neutral superfluid, $T_{SF} = |m^2|^\psi$, which is governed by the shift exponent $\psi = z/(d + z - 2) = 1/2$ for $d \geq 3$. The temperature-dependent line of first-order transitions is given by the condition that the thermal mass, defined by

$$m_T^2 = m^2 - m_c^2 + (q^2/4)T_c^2$$

vanishes. This yields

$$T_c = \frac{2}{q}\sqrt{|g|},$$

where $g = m^2 - m_c^2$ and the critical mass $m_c^2 = (3q^4/32\pi^2) < \varphi >^2$, as given in Eq. 14.15. It is interesting to see in the expression for T_c the charge appearing in the denominator as a genuine dangerously irrelevant interaction. At the quantum critical point of the neutral superfluid, $m^2 = 0$, symmetry has now been broken due to the charge of the particles. This occurs at a critical temperature $T_c = (3/\sqrt{8\pi})q < \varphi >$. The physical origin of the $T = 0$ discontinuous phase transition is the gain in energy due to the expulsion of the electromagnetic field when the system becomes superconducting.

Let us consider in more detail the zero-temperature case where, as shown in the previous chapter, the charged superfluid has a quantum first-order transition at the critical mass m_c. We start examining how the energies of the different ground states that exchange stability at m_c behave in the neighbourhood of the first-order quantum transition. For values of $m > m_c$, the stable ground state for which the effective potential has a global minimum occurs when the order parameter $\varphi_c = 0$, such that $V_{eff}(\varphi_c = 0) = 0$. The value of the effective potential at the metastable minimum $\varphi_c = \langle\varphi\rangle$ is given by

$$V_{eff}^0(\langle\varphi\rangle) = \frac{1}{4}m^2\langle\varphi\rangle^2\left[1 - \frac{m_c^2}{m^2}\right]. \tag{15.5}$$

Then, at $m^2 = m_c^2$ the two ground states at $\varphi_c = 0$ and $\varphi_c = \langle\varphi\rangle$ are degenerate and for $m^2 < m_c^2$, the true ground state is at $\varphi_c = \langle\varphi\rangle$.

The effective potential at $T = 0$ represents the ground state energy (Jackiw, 1973) and close to the critical mass m_c, we find, $V_{eff}^0 \propto |m^2 - m_c^2| \propto |g|^{2-\alpha}$, which implies that the critical exponent $\alpha = 1$. Using hyperscaling we get for the correlation length exponent, $\nu = 1/(d + z)$. This result confirms the value for the exponent α obtained before. As for the correlation length exponent, we will verify the above prediction when we consider the case of $T \neq 0$ below.

The *latent heat* for the transition, i.e. the amount of work required to bring one phase into another is given by

$$L_h = (A_+ + A_-) = \frac{1}{4}m_c^2 \langle\varphi\rangle^2$$

where we used $A_+ = 0$.

We now consider the finite temperature case. We imagine the system tuned at the critical value of the mass for the first-order quantum phase transition, i.e. with the mass fixed at $m^2 = m_c^2$ where $m_c^2 = (3q^4/32\pi^2) < \varphi >^2$. We start at high temperatures, $T \gg \sqrt{m_c^2 + q^2} < \varphi >^2$, and decrease it along the line pointed by the arrow in Fig. 15.1. In this temperature range, which corresponds to the scaling regime I of Fig. 15.1, the function $I_d(y \to 0)$ saturates, for example, for $d = 3$, $I_3(y < 0.12) \approx -2.16$. In this case the effective potential,

$$V_{eff}(T) \approx \frac{1}{4}m^2 \langle\varphi\rangle^2 |g| \left\{ 1 - \frac{4.32}{\pi^2 m^2 \langle\varphi\rangle^2} \frac{T^{d+1}}{|g|} \right\}$$

can be cast in the scaling form,

$$V_{eff}(T) \propto |g|^{2-\alpha} F\left[\frac{T}{T_\times} \right],$$

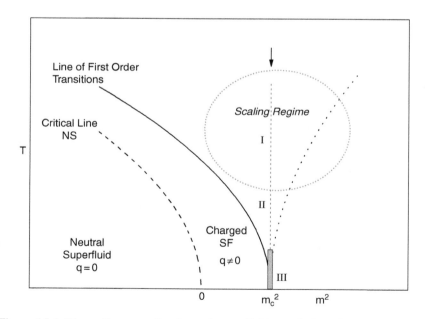

Figure 15.1 Phase diagram of a charged superfluid coupled to photons. For completeness we show also the critical line of the neutral superfluid. Along the trajectory $m^2 = m_c^2$ one can distinguish different regimes as explained in the text.

with $F(0) = constant$. This scaling form is reminiscent of that for the free energy close to a quantum critical point. In the present case of a discontinuous zero temperature transition, the critical exponent $\alpha = 1$ and the characteristic temperature is

$$T_\times \propto |g|^{\nu z} \propto |g|^{\frac{z}{d+z}} = |g|^{\frac{1}{d+1}} = |g|^{\frac{1}{4}},$$

with $\nu = 1/(d + z)$ and the dynamic exponent $z = 1$. This is consistent with the previous scaling prediction for the correlation length exponent.

In the scaling regime, the free energy density has the scaling form $f(m = m_c, T) \propto T^{(d+z)/z}$ and the specific heat is given by

$$C/T\big|_{(m=m_c,T)} \propto T^{\frac{d-z}{z}}. \tag{15.6}$$

Then the thermodynamic behaviour along the line $m^2 = m_c^2$, $T \to 0$ in the scaling regime I is the same as when approaching the quantum critical point of the neutral superfluid, along the critical trajectory $m^2 = 0$, $T \to 0$. The system is unaware of the change in the nature of the zero temperature transition and at such high temperatures, charge is irrelevant. The critical exponents that determine the scaling behaviour of thermodynamic quantities in regime I are the same as those associated with the QCP of the *uncoupled system*, which in the present case is the neutral superfluid.

When further decreasing temperature along the line $m^2 = m_c^2$ there is an intermediate, *non-universal* regime (regime II in Fig. 15.1). For $d = 3$ and $T \sim \sqrt{m_c^2 + q^2 < \varphi >^2}$ the specific heat $C/T^{d/z} \propto \ln T$ as can be easily shown by integrating numerically the function $I_3(y)$.

Finally, at very low temperatures, for $T \ll \sqrt{m_c^2 + q^2 < \varphi >^2}$ the system reaches regime III of Fig. 15.1. In this regime, the specific heat vanishes exponentially with temperature, $C/T^{d/z} \propto \exp(-\Delta/T)$. The gap for thermal excitations is given by $\Delta = \sqrt{m_c^2 + q^2 < \varphi >^2} \approx \sqrt{q^2 < \varphi >^2} = 1/(\sqrt{2}\xi_L)$. The correlation length which grows along the line $m^2 = m_c^2$ with decreasing temperature reaches saturation in regime III at a value $\xi = \xi_S = \sqrt{2}\xi_L$. Let us recall the two characteristic lengths of this problem, the correlation length given by, $\xi = 1/(\sqrt{2m^2})$ and the London penetration depth $\xi_L = 1/\sqrt{2q^2| < \varphi > |^2}$. Notice that the saturation of the correlation length occurs at the critical value of the Landau–Ginzburg parameter $\kappa = \xi_L/\xi = 1/\sqrt{2}$, separating type I and type II superconductors.

The exponential dependence of the specific heat can be attributed to gapped excitations inside superconducting bubbles of finite size $L \sim \xi_S$, which start to nucleate above the quantum first-order transition. The gap between the states in these bubbles is $\Delta \sim \xi_S^{-z} = (\sqrt{2}\xi_L)^{-z}$, with the dynamic exponent $z = 1$. The length scale of the bubbles is essentially fixed by the London penetration depth ξ_L.

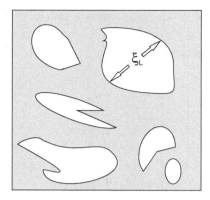

Figure 15.2 When the temperature is decreased at $m = m_c$ bubbles of character-istic length ξ_L begin to form. Excitations in these bubbles of finite size give rise to an exponential dependence of the specific heat at low temperatures.

Although the results above have been obtained for a particular model, we expect that for any fluctuation induced weak first-order quantum phase transition, the behaviour in the scaling regime I and III should be universal. This universality is to be understood in the sense that, for the system at the critical point where the quantum first-order transition occurs and finite temperatures:

- For sufficiently high temperatures it will obey scaling with the dynamic expo-nent the same as that of the uncoupled system. The correlation length exponent takes the value $\nu = 1/(d + z)$.
- At the lowest temperatures the correlation length saturates at a value ξ_S; that is, of the order of the other length scale in the problem. The thermodynamic behaviour is thermally activated with a gap for excitations $\Delta \sim (\xi_S)^{-z}$.

The arguments that yield $\alpha = 1$ and $\nu = 1/(d + z)$ are sufficiently general that we can expect that these results are valid for any discontinuous quantum tran-sition with a *latent heat* and not only for weak or fluctuation-induced first-order transitions.

We can check these scaling predictions for discontinuous quantum transitions by comparing with results on systems where exact solutions are known.

As an example, let us consider the biquadratic spin-1 chain (Barber and Batchelor, 1989), described by the following Hamiltonian:

$$H = -\sum_i \epsilon_i (\vec{S}_i \cdot \vec{S}_{i+1})^2 \tag{15.7}$$

with

$$\epsilon_i = \begin{cases} 1 & \text{if } i \text{ is odd} \\ \lambda & \text{if } i \text{ is even.} \end{cases} \tag{15.8}$$

At $\lambda = 1$ there is a zero-temperature first-order phase transition where two spontaneously dimerised ground states exchange stability (Barber and Batchelor, 1989).
The ground state energy can be written as

$$E_g(\lambda) - E_g(\lambda = 1) = A_{\pm}|1 - \lambda| \tag{15.9}$$

consistent with $\alpha = 1$ and the latent heat $L = -(A_+ + A_-)$ can be exactly
obtained (Barber and Batchelor, 1989). Furthermore, in this case the correlation length exponent has been directly obtained from finite lattice calculations (Sólyom, 1987). The numerical value, $\nu \approx 0.5$ agrees with the expected
value $\nu = 1/(d + z) = 1/2$ since for this transition $z = 1$ (Sólyom, 1987). Other
examples can be found in the literature confirming the scaling results (Ribeiro
et al., 2011).

15.3 Conclusions

In this chapter we presented a scaling approach for quantum first-order phase transitions. In spite of the discontinuous nature of these transitions, scaling concepts
turn out to be useful and powerful. In particular, this is the case for fluctuation-
induced weak first-order transitions, whenever fluctuations are responsible for
changing the nature of the transition from second to weak first order. We have studied in detail the special case of a charged superfluid whose quasi-particles couple
to fluctuations of the electromagnetic field. The results that should hold in general
show that as the system approaches the discontinuous transition from non-zero
temperature, we can distinguish three different regimes. A scaling, high temperature regime, with critical exponents that we have determined. In particular the
dynamic exponent coincides with that of the uncoupled system. Further decreasing
the temperature there is a non-universal regime which may depend on the particular dynamics of the original system and on the nature of the additional modes.
Finally, at the lowest temperatures there is a regime which is characteristic of the
first-order nature of the zero-temperature transition. At such low temperatures the
correlation length saturates at a finite value. The thermodynamic properties have a
contribution which is thermally activated. This can be traced to excitations inside
bubbles of the incipient phase and that are gapped due to the finite size of these
bubbles.

Appendix

A.1 Green's Functions

The time-dependence of an operator in the Heisenberg representation is given by

$$A(t) = e^{iHt} A e^{-iHt},$$

where $H = H_0 - \mu N$, with μ the chemical potential. Its equation of motion is given by

$$i\frac{dA}{dt} = [A(t), H].$$

In the grand canonical ensemble the thermodynamical average is given by

$$< A > = \frac{Tr e^{-\beta H} A}{Tr e^{-\beta H}} = Tr\rho A,$$

where $\rho = e^{-\beta H} / Tr e^{-\beta H}$ is the density matrix.

Let us define retarded and advanced Green's functions (Tyablikov, 1967), respectively,

$$G^r_{AB}(t, t') = << A(t); B(t') >>^r = \theta(t - t') < [A(t), B(t')]_\eta >$$
$$G^a_{AB}(t, t') = << A(t); B(t') >>^a = -\theta(t' - t) < [A(t), B(t')]_\eta >$$

where $[A, B]_\eta = AB - \eta BA$ with $\eta = -1$ for fermionic and $\eta = +1$ for bosonic operators and $\theta(t)$ is the step function.

These Green's functions obey the following equation of motion,

$$i\frac{d}{dt} << A(t); B(t') >> = i\delta(t - t') < [A(t), B(t')]_\eta > + << [A(t), H]; B(t') >> .$$

(A.1)

In general the last term generates higher-order Green's functions. We used above that $d\theta(t)/dt = \delta(t)$.

Introducing the Fourier transform of the time-dependent Green's functions

$$G_{AB}(\omega) =<< A; B >>_\omega= \frac{1}{2\pi} \int_{-\infty}^{+\infty} dt\, G_{AB}(t) e^{i\omega t},$$

the equation of motion can be written as

$$\omega << A; B >>_\omega= \frac{i}{2\pi} < [A(t), B(t')]_\eta > + << [A, H]; B >>_\omega .$$

Spectral Representation of the Green's Functions

The spectral density $I_{BA}(\omega)$ is defined by

$$I_{BA}(\omega) = \frac{1}{Z} \sum_{m,n} e^{-\beta E_n} [B_{nm} A_{mn} \delta(\omega + \omega_{nm}))], \tag{A.2}$$

where $A_{nm} =< n|A|m >$, $\omega_{nm} = E_n - E_m$ with $|n >$ and E_n, the eigenstates and eigenvalues of the Hamiltonian H. Also $Z = Tr e^{-\beta H}$.

It is easy to show that the time-dependent correlations functions can be written as

$$< B(t)A(t') >= \int_{-\infty}^{+\infty} d\omega I_{BA}(\omega) e^{-i\omega(t-t')}$$
$$< A(t)B(t') >= \int_{-\infty}^{+\infty} d\omega I_{BA}(\omega) e^{\beta\omega} e^{-i\omega(t-t')},$$

such that

$$< [A(t), B(t')]_\eta >= \int_{-\infty}^{+\infty} d\omega I_{BA}(\omega)(e^{\beta\omega} - \eta) e^{-i\omega(t-t')}$$

and

$$G_{AB}^r(\omega) = \frac{1}{2\pi} \int_{-\infty}^{+\infty} d\omega' I_{BA}(\omega')(e^{\beta\omega'} - \eta) \int_{-\infty}^{\infty} dt e^{i(\omega-\omega')t} \theta(t).$$

From these equations we can obtain the spectral representations of the retarded and advanced Green's functions:

$$G_{AB}^r(\omega) = \frac{1}{2\pi} \int_{-\infty}^{+\infty} \frac{d\omega'}{\omega-\omega'+i\epsilon} I_{BA}(\omega')(e^{\beta\omega'} - \eta)$$
$$G_{AB}^a(\omega) = \frac{1}{2\pi} \int_{-\infty}^{+\infty} \frac{d\omega'}{\omega-\omega'-i\epsilon} I_{BA}(\omega')(e^{\beta\omega'} - \eta).$$

It is clear from above that it is useful to abandon the indexes r and a, and make an analytic continuation of $G_{AB}(\omega)$, such that

$$G_{AB}(\omega+i\epsilon)-G_{AB}(\omega-i\epsilon) = \frac{1}{2\pi} \int_{-\infty}^{+\infty} d\omega' I_{BA}(\omega')(e^{\beta\omega'} - \eta)[\frac{1}{\omega-\omega'+i\epsilon} - \frac{1}{\omega-\omega'-i\epsilon}]$$
$$= I_{BA}(\omega)(e^{\beta\omega} - \eta)$$

or

$$I_{BA}(\omega) = \frac{1}{(e^{\beta\omega} - \eta)}[G_{AB}(\omega + i\epsilon) - G_{AB}(\omega - i\epsilon)].$$

Taking into account the **k**-dependence of the Green's functions we get

$$< B(0)A(0) > = \sum_{\mathbf{k}} \int_{-\infty}^{+\infty} d\omega I_{BA}(\mathbf{k}, \omega)$$

$$= \sum_{\mathbf{k}} \int_{-\infty}^{+\infty} \frac{d\omega}{e^{\beta\omega} - \eta}[G_{AB}(\mathbf{k}, \omega + i\epsilon) - G_{AB}(\mathbf{k}, \omega - i\epsilon)],$$

$$(A.3)$$

which allows us to obtain a correlation function from the appropriate Green's function. This is essentially a modified version of the fluctuation–dissipation theorem.

In general, the Green's functions we deal here are of two forms:

- In the simplest case

$$G_{AB}(\mathbf{k}, \omega) = \frac{i}{2\pi} \frac{D(\mathbf{k}, \omega)}{\omega - \epsilon_k} \tag{A.4}$$

such that,

$$G_{AB}(\mathbf{k}, \omega + i\epsilon) - G_{AB}(\mathbf{k}, \omega - i\epsilon) = D(\mathbf{k}, \omega)\delta(\omega - \epsilon_k).$$

- Also

$$G_{AB}(\mathbf{k}, \omega) = \frac{i}{2\pi} \frac{D(\mathbf{k}, \omega)}{E(\mathbf{k}, \omega)}. \tag{A.5}$$

For simplicity, we assume $E(\mathbf{k}, \omega)$ is of second degree in ω with roots $\omega_1(k)$ and $\omega_2(k)$. In this case, we get

$$G_{AB}(\mathbf{k}, \omega + i\epsilon) - G_{AB}(\mathbf{k}, \omega - i\epsilon) = \frac{D(\mathbf{k}, \omega)}{\omega_1(k) - \omega_2(k)}[\delta(\omega - \omega_1(k)) - \delta(\omega - \omega_2(k))].$$

This procedure can easily be extended for any number of roots of the denominator $E(\mathbf{k}, \omega)$.

References

Alexandrov, Victor and Coleman, Piers (2014), *Phys. Rev.* **B 90**, 115147.

Alicea, Jason (2012), *Reports on Progress in Physics*, **75**, 076501.

Anderson, P. W. (1963), *Phys. Rev.* **130**, 439.

Anderson, P. W. (1970), *J. Phys. C: Sol. St. Phys.* **3**, 2436.

Aubry, S and André, G. (1980), *Ann. Israel Phys. Soc.* **3**, 133.

Auerbach, A. (1994), *Interacting Electrons and Quantum Magnetism*, New York: Springer-Verlag.

Barber, M. and Batchelor, M. T. (1989) *Phys. Rev.* **B40**, 4621.

Bauer, E. D., Yang, Yi-feng, Capan, C., Urbano, R. R., Miclea, C. F., Sakai, H., Ronning, F., Graf, M. J., Balatsky, A. V., Movshovich, R., Bianchi, A. D., Reyes, A. P., Kuhns, P. L., Thompson, J. D. and Fisk, Z. (2011) *PNAS* **108**, 6857.

Beal-Monod, M. T., Ma, S. K. and Fradkin, D. R. (1968), *Phys. Rev. Lett.* **20**, 929.

Belitz, D. and Kirkpatrick, T. R. (1994), *Rev. Mod. Phys.* **66**, 361.

Belitz, D., Kirkpatrick, T. R. and Vojta, T. (2005), *Rev. Mod. Phys.* **77**, 579.

Belitz, D., Kirkpatrick, T. R. and Rollbüler, J. (2005), *Phys. Rev. Lett.* **94**, 247205.

Belvedere, L. (1999), *Lecture Notes*, *Universidade Federal Fluminense*, unpublished.

Blanter, Ya. M., Kaganov, M. I., Pantsulaya, A. V., Varlamov, A. A. (1994), *Phys. Rep.* **245**, 159.

Blundell, Stephen (2001), *Magnetism in Condensed Matter* (Oxford Master Series in Condensed Matter Physics).

Boechat, B., dos Santos, R. R. and Continentino, M. A. (1994), *Phys. Rev.* **B49**, R6404.

Bray, A. J. and Moore, M. A. (1985), *J. Phys. C: Solid State Phys.* **18**, L927.

Brézin, E. (2015), in *Ken Wilson Memorial Volume: Renormalisation, Lattice Gauge Theory, the Operator Product Expansion and Quantum Fields*, Baaquie, B. E., Huang, K., Peskin, M. E. and Phua, K. K. (eds.), World Scientific Publishing Co. Pte. Ltd.

Brinkman, W. F. and Rice, T. M. (1970), *Phys. Rev.* **B2**, 4302.

Büttiker, M., Imry, Y., and Landauer, R. (1983), *Phys. Lett.* **A 96**, 365.

Byers, N. and Yang, C. N. (1961), *Phys. Rev. Lett.* **7**, 46.

Caldas, H. and Continentino, M. A. (2012), *Phys. Rev.* **B 86**, 144503.

Cardy, J. (1996), *Scaling and Renormalization in Statistical Physics*, Cambridge University Press.

Castro Neto, A. H., Castilla, G. E. and Jones, B. A. (1998), *Phys. Rev. Lett.* **81**, 3531.

Cestari, J. C. C., Foerster, A. and Gusmão M. A. (2010), *Phys. Rev.* **A 82**, 063634.

Cestari, J. C. C., Foerster, A. Gusmão, M. A. and Continentino, M. A. (2011), *Phys. Rev.* **A 84**, 055601.

Chakravarty, S., Halperin, B. I. and Nelson, D. (1988), *Phys. Rev. Lett.* **60**, 1057.

Chandrasekhar, B. S. (1962) *Appl. Phys. Lett.* **1**, 7.

Clogston, A. M. (1962) *Phys. Rev. Lett.* **9**, 266.

Coleman, P. (1983), *Phys. Rev.* **B28**, 5255.

Coleman, S. and Weinberg, E. (1973), *Phys. Rev.* **D7**, 1888.

Coleman, Sidney (1988), *Aspects of Symmetry*, Cambridge University Press.

Coqblin, B. (1977), *Electronic Structure of Rare-Earth Metals and Alloys: the Magnetic Heavy Rare-Earths*, London: Academic Press Inc.

Combescot, R. and Mora, C. (2005), *Eur. Phys. J.* **B 44**, 189.

Continentino, M. A. and Rivier, N. (1977), *J. Phys. C: Solid State Physics* **10**, 3613.

Continentino, M. A., Japiassu, G. and Troper, A. (1989) *Phys. Rev.* **B39**, 9734.

Continentino, M. A. (1991a), *J. de Physique I* **1**, 693.

Continentino, M. A. (1991b), *Phys. Rev.* **B43**, 6292.

Continentino, M. A. (1993), *Phys. Rev.* **B47**, 11587.

Continentino, M. A. (1994a), *Phys. Rep.*, **239**, 179.

Continentino, M. A. and Coutinho-Filho, M. D. (1994b), *Sol. St. Comm.* **90**, 619.

Continentino, M. A. (1995a), *Phys. Lett.* **A 197**, 417.

Continentino, M. A., Elschner, B. and Jakob, G. (1995b), *Europhys. Lett.* **31**, 485.

Continentino, M. A. (1996), *Zeitschrift für Physik* **B101**, 197.

Continentino, M. A. (1998), *Phys. Rev.* **B57**, 5966.

Continentino, M. A. (2000), *Eur. Phys. J.* **B13**, 31.

Continentino, M. A. and Ferreira, A. S. (2004), *PHYSICA A* **339**, 461.

Continentino, M. A. and Ferreira, A. S. (2007), *JMMM*, **310**, 828.

Continentino, M. A. (2011), *Braz. J. Phys.* **41**, 201.

Continentino, M. A., Deus, F. and Caldas, H. (2014) *Phys. Lett.* **A 378**, 1561.

Continentino, M. A., Caldas, H., Nozadze, D., Trivedi, N. (2014), *Phys. Lett.* **A378**, 3340.

DeGottardi, W., Sen, Diptiman and Vishveshwara, S. (2011), *New Journal of Physics* **13**, 065028.

de Mello, E. V and Continentino, M. A. (1990), *J.Phys. Condensed Matter* **2**, 4161.

Deus, Fernanda and Continentino, Mucio (2013), *Philo. Mag.* **93**, 3062.

Dolan, L. and Jackiw, R. (1974), *Phys. Rev.* **D9**, 3320.

Doniach, S. (1977), *Physica* **B91**, 231;

dos Santos, R. R. (1982), *J. Phys. C: Solid State Phys.* **15**, 3141.

Enderlein, C., Ramos, S. M., Bittencourt, M., Continentino, M. A., Brewer, W. and Baggio-Saitovich, E. (2013), *J. Appl. Phys.* **114**, 143711.

Enderlein, C., Fontes, M., Baggio-Saitovitch, E. and Continentino, M. A. (2016), *Journal of Magnetism and Magnetic Materials* **398**, 270.

Ferreira, A. S., Continentino, M. A. and Marino, E. C. (2004), *Phys. Rev.* **B70**, 174507.

Fisher, M. E. and Berker, A. N. (1982), *Phys. Rev.* **B26**, 2507.

Fisher, M. E. (1983), in *Critical Phenomena*, ed. by Hahne, F. J. W. (ed.) Berlin: Springer Verlag, 1.

Fisher, M. P. A., Weichman, P. B., Grinstein, G. and Fisher, D. S. (1989), *Phys. Rev.* **B40**, 546.

Fisher, M. P. A., Grinstein, G. and Girwin, S. (1990) *Phys. Rev. Lett.* **64**, 587.

Kambe, S., Flouquet, J. and Hargreaves T. E. (1997), *J. Low Temp. Phys.* **108**, 383.

Foo, E-Ni and Giangiulio, D. J. (1976), *PHYSICA* **84B**, 167.

Fradkin, E. (1991), *Field Theories of Condensed Matter Systems*, Redwood City, CA: Addison-Wesley.

Freitas, D. C., Rodière, P., Núñez, M., Garbarino, G., Sulpice, A., Marcus, J., Gay, F., Continentino, M. A. and Núñez-Regueiro, M. (2015), *Phys. Rev.* **B 92**, 205123.

Fulde, P. and Ferrell, R.A. (1964), *Phys. Rev.* **A135**, 550.

Georges, A., Kotliar, G., Krauth, W. and Rozenberg, M. J. (1996), *Rev. Mod. Phys.* **68**, 73.

Georges, A., Kotliar, G., Krauth, W. and Rozenberg, M. J. (1996), *Rev. Mod. Phys.* **68**, 13.

Goldenfeld, N. (1992), *Lectures on Phase Transitions and the Renormalization Group*, USA: Addison-Wesley Publishing Company.

Gorkov, L. P. and Melik-Barkhudarov, T. K. (1961), *Zh. Eksp. Teor. Fiz.* **40**, 1452; (1961), *Sov. Phys. JETP* **13**, 1018).

Granato, E. and Continentino, M. A. (1993), *Phys. Rev.* **B**, 15977.

Gutzwiller, M. C. (1965), *Phys. Rev.* **A137**, A1726.

Halperin, B. I. and Rice, T. M. (1968), *Rev. Mod. Phys.* **40**, 755.

Halperin, B. I., Lubensky, T. and Ma, S.-K. (1974), *Phys. Rev. Lett.* **32**, 292.

Harper, P. G. (1955), *Proc. Phys. Soc. Lond. Sect.* **A 68**, 874.

Hasan, M. Z. and Kane, C. L. (2010) *Rev. Mod. Phys.* **82**, 3045.

Hayden, S. M. *et al.* (2000), *Phys. Rev. Lett.* **84**, 999.

Hertz, J. A. (1976), *Phys. Rev.* **B14**, 1165.

Hiramoto, H. and Kohmoto M. (1989), *Phys. Rev.* **B 40**, 8225.

Hofstadter, D. R. (1976), *Phys. Rev.* **B 14**, 2239.

Huang, K. (2015), in *Ken Wilson Memorial Volume: Renormalization, Lattice Gauge Theory, the Operator Product Expansion and Quantum Fields*, Baaquie, B. E., Huang, K., Peskin, M. E. and Phua, K. K. (eds.), *World Scientific*, 19.

Hubbard, J. (1963), *Proc. Roy. Soc. London* **Ser. A276**, 238.

Hubbard, J. (1964a), *Proc. Roy. Soc. London* **Ser. A277**, 237.

Hubbard, J. (1964b), *Proc. Roy. Soc. London* **Ser. A281**, 401.

Imada, M., Fujimori, A. and Tokura, Y. (1998), *Rev. Mod. Phys.* **70**, 1039.

Ingold, G-L., Wobst, A., Aulbach, C. and Hanggi, P. (2002), *Eur. Phys. J.* **B 30**, 175.

Jaccard, D., Behnia, K. and Sierro, J. (1992), *Phys. Lett.* **A 163**, 475.

Jackiw, R. (1973), *Phys. Rev.* **D9**, 1686.

Jullien, R. (1981), *Can. J. Phys.* **59**, 605.

Kadanoff, L. P., Gotze, W., Hamblen, D. *et al.* (1967), *Rev. Mod. Phys.* **39**, 395.

Kadowaki, K. and Woods, S. B. (1986), *Solid St. Comm.* **58**, 507.

Kaganov, M. I. and Möbius A. (1984), *Sov. Phys. JETP* **59**, 405.

Kanamori, J. (1963), *Prog. Theor. Phys.* **30**, 275.

Kawakami, N. and Yang, S.-K. (1990), *Phys. Rev. Lett.* **65**, 3063.

Kehrein, Stefan (2006) in *The Flow Equation Approach to Many-Particle Systems*, Springer Tracts in Modern Physics Volume 217, Berlin Heidelberg: Springer-Verlag.

Kitaev, A. Y. (2001), *Physics-Uspekhi* **44**, 131.

Kitaev, A. Y. (2003), *Annals of Physics* **303**, 2.

Kleinert, H. (1987), *Gauge Fields in Condensed Matter/Volume 1: Superflow and Vortex Lines/Volume 2: Stresses and Defects*, World Scientific Publishing Company, Incorporated.

Kohmoto, M. (1983), *Phys. Rev. Lett* **51**, 1198.

Kohn, W. (1964), *Phys. Rev.* **A133**, 171.

Kotliar, G. and Ruckenstein, A. (1986), *Phys. Rev. Lett.* **57**, 11.

Lacroix, C. and Pinettes, C. (1995), *Physica B* **206&207**, 11.

Landau, L. D. and Lifshitz, E. M. (1980), *Statistical Mechanics*, Oxford: Pergamon Press.

Larkin, A. I. and Ovchinnikov, Yu. N. (1965) *Sov. Phys. JETP* **20**, 762.

Lavagna, M. (1990), *Phys. Rev.* **B41**, 142.

Lederer, P. and Mills, D. L.(1968), *Phys. Rev.* **165**, 837.

Lee, P. A. and Ramakrishnan, T. V. (1985), *Rev. Mod. Phys.* **57**, 287.

Linde, A. D. (1979), *Rep. Prog. Phys.* **42**, 389.

Löhneysen, v. H. *et al.* (1994), *Phys. Rev. Lett.* **72**, 3262.

Loram, J. W., Mirza, K. A. and Freeman, P. F. (1990) *Physica* **C 171**, 243.

Luengo, C. A., Maple, M. B. and Fertig, W. A. (1972) *Solid State Commun.* **11**, 1445.

Ma, S.-K. (1976), *Modern Theory of Critical Phenomena*, Reading, MA: Addison-Wesley.

Malbuisson, A. P. C., Nogueira, F. S. and Svaiter, N. F. (1996), *Mod. Phys. Lett.* **11**, 749.

Mathur, N. D., Grosche, F. M., Julian, S. R., Walker, I. R., Freye, D. M., Haselwimmer, R. K. W. and Lonzarich, G. G. (1998), *Nature* **394**, 39.

Medeiros, S. N. *et al.* (2000), *Physica* **B281&282**, 340.

Millis, A. J., Monien, H. and Pines, D. (1990), *Phys. Rev.* **B 42**, 167.

Millis, A. J. (1993), *Phys. Rev.* **B48**, 7183.

Mineev, V. P. and Sigrist, M. (2001), *Phys. Rev.* **B 63**, 172504.

Mineev, V. P. and Zhitomirsky, M. E. (2005) *Phys. Rev.* **B72**, 014432.

Miranda, E., Dobrosavljevic, V. and Kotliar, G. (1996), *Phys. Rev. Lett.* **78**, 290; Miranda, E., García, D. J., Hallberg, K. and Rozenberg, M. J. (2008), *Physica* **B 403**, 1465.

Mitra, A. (2008), *Phys. Rev.* **B 78**, 214512.

Moriya, T. (1985) *Spin Fluctuations in Itinerant Electron Magnetism*, Berlin Heidelberg: Springer-Verlag.

Moriya, T. and Takimoto, T. (1995), *J. Phys. Soc. Jpn.* **64**, 960.

Mott, N. F. (1974), *Metal-Insulator Transitions*, London: Taylor and Francis.

Murakami, Shuichi (2007), *New Journal of Physics* **9**, 356.

Negele, J. W. and Orland, H. (1988), *Quantum Many-Particle Systems*, Redwood City, CA: Addison-Wesley.

Nienhuis, B. and Naunberg, M. (1975), *Phys. Rev. Lett.* **35**, 477.

Nogueira, F. S. (1996), *PhD. Thesis*, Centro Brasileiro de Pesquisas Fisicas, unpublished.

Nozadze, D. and Trivedi, N. (2016) *Phys. Rev.* **B 93**, 064512.

Nozières, P. (1986), *Magnétisme et Localization dans les Liquides de Fermi*, Collège de France, unpublished.

Oliveira, P. M., Continentino, M. A. and Anda, E. (1984), *Phys. Rev.* **B29**, 2808.

Padilha, Igor T. and Continentino, M. A. (2009), *Physica B – Condensed Matter*, **404**, 2920.

Peskin, M. E. and Schroeder, D. V. (1995), *An Introduction to Quantum Field Theory*, First Edition, Frontiers in Physics, Westview Press.

Pfeuty, P. (1970), *Ann. Phys.* **57**, 79.

Pfeuty, P., Jullien, R. and Penson, K. A. (1982), *Real Space Renormalisation Group*, Berlin: Springer Verlag, 119.

Pfeuty, P., Jasnow, D. and Fisher, M. E. (1974), *Phys. Rev.* **B10**, 2088.

Prokovsky, V. L. and Tapalov, A. L. (1979), *Phys. Rev. Lett.* **41**, 65.

Puel, T. O., Sacramento, P. D. and Continentino, M. A. (2015), *J. of Phys. – Cond. Mat.* **27**, 422002.

Ramazashvili, R. (1999), *Phys. Rev.* **B60**, 7314.

Ramazashvili, R. and Coleman, P. (1997), *Phys. Rev. Lett.* **79**, 3752.

Ramires, A. and Continentino, M. A. (2010), *J. Phys.: Condens. Matter* **22**, 485701.

Rezende, S. M. (1975) in *Brazilian Symposium on Theoretical Physics, Livros Técnicos e Científicos Editora S.A.*, edited by Erasmo Ferreira and Rio de Janeiro, RJ, Vol. 2, 3; Rezende, S. M., King, A. R., White, R. M. and Timbie, J. P. (1977), *Phys. Rev.* **B 16**, 1126.

Ribeiro, F. G., de Lima J. P. and Gonçalves, L. L. (2011), *JMMM* **323**, 39.

Rosch, A., Schröder, A. Stockert, O. and Löhneysen, H. (1997), *Phys. Rev. Lett.* **79**, 159.

Sachdev, S, (1996), *Proceedings of the 19th IUPAP International Conference on Statistical Mechanics*, Singapore: World Scientific.

Sachdev, S. (1999), *Quantum Phase Transitions*, Cambridge University Press.

Sachdev, S., Read, N. and Oppermann, R. (1995), *Phys. Rev.* **B52**, 10286.

Sachdev, S. (1994), *Z. Phys.* **B94**, 469.

Sachdev, S. (2011), *Quantum Phase Transitions*, Second Edition, Cambridge University Press.

Samokhin, K. V. and Marénko, M. S. (2006), *Phys. Rev.* **B73**, 144502.

Scalapino, D. J., White, S. R. and Zhang, S. C. (1992), *Phys. Rev. Lett.* **68**, 2830.

Scalapino, D. J., White, S. R. and Zhang, S. C. (1993), *Phys. Rev.* **B 47**, 7995.

Schröder, A., Aeppli, G., Bucher, E. *et al.* (1998), *Phys. Rev. Lett.* **80**, 5623.

Schuberth, E., Tippmann, M., Steinke, L., Lausberg, S., Steppke, A., Brando, M., Krellner, C., Geibel, C., Yu, Rong, Si, Qimiao, Steglich and F. (2106), *SCIENCE* **351**, 485.

Shastry, B. and Sutherland, B. (1990), *Phys. Rev. Lett.* **65**, 243.

Shimahara, Hiroshi (1994), *Phys. Rev.* **B 50**, 12760.

Shimahara, Hiroshi (1998), *J. Phys. Soc. Jpn.* **67**, 736.

Shockley, W. (1939) *Phys. Rev.* **56**, 317.

Sólyom, J. (1987), *Phys. Rev.* **B36**, 8642

Sondhi, S. L. *et al.* (1997), *Rev. Mod. Phys.* **69**, 315.

Stanley, H. E. (1971), *Introduction to Phase Transitions and Critical Phenomena*, Oxford University Press.

Steglich, F., Geibel, C., Helfrisch, R. *et al.* (1998), *J. Phys. Chem. Solids* **59**, 2190.

Steglich, F., Geibel, C., Modler, R. *et al.* (1995), *J. Low Temp. Phys.* **99**, 267.

Strogatz, S. (2014), *Nonlinear Dynamics and Chaos: With Applications to Physics, Biology, Chemistry, and Engineering (Studies in Nonlinearity)*, Second Edition, USA: Westview Press.

Suhl, H., Matthias, B. T. and Walker, L. R. (1959), *Phys. Rev. Lett.* **3**, 552.

Takimoto, T. and Moriya, T. (1996), *Solid State Comm.* **99**, 457.

Thompson, J. D. and Lawrence, J. M. (1994), *Handbook on the Physics and Chemistry of Rare Earths, Lanthanides/Actinides: Physics-II, K. A. Gschneider Jr., L. Eyring, G. H. Lander and G. R. Choppin* (eds), Elsevier Science B.V., Chapter 133 (19), 383.

Thouless, D. J. (1960), *Ann. Phys.* **10**, 553.

Toulouse, G. (1974), *Nuovo Cimento* **B23**, 234.

Toulouse, G. and Pfeuty, P (1975), *Introduction au Groupe de Renormalisation et à ses Applications*, Presses Universitaires de Grenoble.

Tyablikov S. V. (1967) *Methods in the Quantum Theory of Magnetism*, New York, NY: Plenum Press.

Vitoriano, C., Bejan, L. B., Macedo, A. M. S. and Coutinho-Filho, M. D. (2000), *Phys. Rev.* **61**, 7941.

Vollhardt, D. (1984), *Rev. Mod. Phys.* **56**, 99.

Vollhardt. D. (2012), *Ann. Phys.* (Berlin) **524**, 1.

Volovik, G. E. (2009), *The Universe in a Helium Droplet*, New York, NY: Oxford University Press.

Vučičević, J., Tanasković, D., Rozenberg, M. J. and Dobrosavljević, V. (2015), *Phys. Rev. Lett.* **114**, 246402.

Wilson, K. G. (1971), *Phys. Rev.* **B4**, 3174.

Wilson, K. G. (1975), *Rev. Mod. Phys.* **47**, 1773.

White, S. R. (1992), *Phys. Rev. Lett.* **69**, 2863.

White, S. R. (1993), *Phys. Rev.* **B 48**, 10345.

Young, A. P. (1975), *J. Phys. C: Solid State Phys.* **8**, L309.

Yu, Rong, Miclea, Corneliu F., Weickert, F., Movshovich, R., Paduan-Filho, A., Zapf, V. S. and Roscilde, T. (2012) *Phys. Rev.* **B 86**, 134421.

Index